Data Science for Sustainable Development Goals

This book presents real-life applications of data science, artificial intelligence, and big data for the Sustainable Development Goals. It includes a list of case studies from different states across India. The case studies in the book use both structured and unstructured data like numeric data, satellite data, online search, data captured through drones, textual data, and video/image data across the various chapters for analysis. It explores various aspects of data science starting from on ground data collection to dashboard based reporting, unsupervised methods like clustering for grouping data points, use of artificial intelligence, machine learning and deep learning, search or information retrieval, time series forecasting, and optimization.

- It showcases data science decision-making processes, driving innovation, and solving complex problems in real-life scenarios across sectors like governance, education, healthcare, agriculture and sanitation.
- The SDGs provide a framework for societal development and well-being for all; the data science and big data interventions in this book are aligned towards mapping the various SDGs.
- Most of the data science use cases and initiative projects covered in this book have been implemented by central or state governments across different states of India.
- Shows how data science intervention can transform the social sector, potentially driving positive change and addressing critical societal challenges.
- Explains the fundamentals of data science theories with case studies, including concepts like classification, regression, predictive analytics, optimization, artificial intelligence, deep learning, and time series forecasting for readers of different disciplines.

It serves as a valuable reference for graduate students, researchers, and scholars seeking to deepen their knowledge and engage with real-life applications of data science. It will also serve as a valuable resource for government officers and policy practitioners, providing a range of cases on the use of data-based methods for improving governance and policy making.

Data Science for Sustainable Development Goals

Indian Case Studies

Edited by
Avik Sarkar and
Bappaditya Mukhopadhyay

CRC Press
Taylor & Francis Group
Boca Raton London New York

CRC Press is an imprint of the
Taylor & Francis Group, an **informa** business
A CHAPMAN & HALL BOOK

Designed cover image: Bappaditya Mukhopadhyay

First edition published 2026
by CRC Press
2385 NW Executive Center Drive, Suite 320, Boca Raton FL 33431

and by CRC Press
4 Park Square, Milton Park, Abingdon, Oxon, OX14 4RN

CRC Press is an imprint of Taylor & Francis Group, LLC

ISBN: 978-1-032-78131-0 (hbk)
ISBN: 978-1-032-78367-3 (pbk)
ISBN: 978-1-003-48753-1 (ebk)

DOI: 10.1201/9781003487531

Typeset in Times
by Apex CoVantage LLC

Dedicated to the data science community in India,
which brings a fresh approach to policy intervention.

Contents

SECTION IV Energy, Water and Sanitation

SECTION V Smart City and Urban Development

Foreword

Prime Minister Shri Narendra Modi's aspirational vision of Viksit Bharat (Developed India) aims to achieve a high-income status for India by the time it celebrates the centenary of its independence in 2047. The roadmap emphasizes comprehensive progress across economic, social, environmental, and governance dimensions at both national and state levels.

Data will play a foundational role in achieving Viksit Bharat, acting as the backbone for evidence-based policymaking and efficient governance. Data helps governments transition from intuition-driven to fact-driven policies, enabling them to identify priority sectors, vulnerable populations, and regional disparities more effectively.

Timely and effective data analytics enable monitoring of literacy rates, disease outbreaks, and healthcare access. Data can aid in poverty alleviation by enabling the design of more efficient welfare schemes based on accurate demographic and income data. Data plays a crucial role in performance monitoring by real-time tracking of government programs and ensuring accountability.

India, with its vast population and extensive digital public infrastructure, has emerged as one of the most data-abundant nations globally. Effective harnessing and strategic utilization of this data by central and state governments is imperative for accelerating socio-economic development. In an increasingly interconnected global landscape, a paradigm shift toward data-driven decision-making is empowering policymakers, investors, researchers, and citizens alike. Leveraging data effectively will be instrumental in enabling India to achieve the vision of Viksit Bharat and capitalize on the collective strength of its citizens.

This book brings together real-world applications of data science and big data in advancing the Sustainable Development Goals (SDGs). This is illustrated through case studies from diverse regions across India. The SDGs provide a universally accepted framework for promoting multi-dimensional societal progress and ensuring the well-being of all.

India's growing expertise in data-driven decision-making reflects its strong commitment to these goals and demonstrates how technology and analytics can serve as powerful enablers of inclusive growth.

The chapters in this volume showcase an emerging governance model – one that blends the vision and resources of government institutions with the analytical rigor of data scientists to deliver public goods effectively. In a country as complex and diverse as India, such collaborations are essential for achieving transformative impact and outcomes. Each case study presented here follows a structured approach, encompassing problem formulation, data collection, and solution design, thereby offering actionable insights for practitioners and policymakers alike.

Contributors to this book represent a wide spectrum of stakeholders, including officials from central and state ministries, academics from universities and policy think tanks, and representatives from private enterprises and non-profit organizations. Their collective efforts, as reflected in this volume, underscore the importance of partnerships in conceptualizing, implementing, and scaling data-driven solutions for sustainable development.

The editors have curated these chapters with the intention of demonstrating how evidence-based decision-making can translate into tangible, positive social impact, setting an example for governance models worldwide.

I expect this book to serve as a valuable resource for policymakers, researchers, and data practitioners alike. It will help them harness the power of data for achieving the national goal of Viksit Bharat by 2047.

Rajiv Kumar
Chairman, Pahle India Foundation, New Delhi
(Former vice chairman, NITI Aayog)

Preface

It was during lunch at a conference on data science and public policy at the Constitution Club of India that the idea of compiling use cases of data science for India's Sustainable Development Goal (SDG) journey was first contemplated. While Avik, at that time, was heading the Data Science division of NITI Aayog, Bappa was deeply involved in teaching data science to graduate students. Three things came together. One, the community that used data science in India was growing. Two, India was characterized by extensive market failure that required both macro- and micro-level policy targeting. Finally, the government was eager to embrace data-driven decision-making to address policy interventions. The effort to compile a volume of exemplary case studies began then.

Around 2019–2020, the editors contacted organizations that included officials from the government or vendors involved in the development of data-based initiatives. The individuals involved in these initiatives viewed this as a platform for sharing their work with the broader community of government officials, researchers, policy practitioners, and data scientists. Around that time, COVID-19 struck, creating some delays in finalizing the edited versions of the chapters.

During that period, NITI Aayog had already undertaken a range of data-driven initiatives. The first Sustainable Development Goal index for India was developed, involving the identification of state-specific indicators based on available data. Other data-led activities involved were ranking and indexing states across various areas, including healthcare, quality of school education, and water management. These indices helped the states obtain quantitative evidence on areas for improvement and inform administrative decisions. The aspirational districts program was taken up by the government, which involved the evaluation and ranking of all the districts across India across seven factors grouped into broad categories (health and nutrition, education, agriculture and water resources, financial inclusion and skill development, and infrastructure) and then selecting the ones at the bottom for a targeted development program for the next five years. Energy and climate change are important areas that are segregated across multiple ministries, leading to NITI Aayog working on long-term energy modeling and forecasting for the nation. The energy modeling activity helped policymakers to take the necessary steps in planning India's future energy roadmap. These initiatives showed a focused approach towards the use of data-based methods in the policy-making and governance process.

While the chapters were being edited and compiled, COVID-19 struck. It delayed the work significantly. However, the inordinate delay did have an advantage. Along with its objective of creating a framework for SDG, NITI Aayog developed a strategy and action plan for India's development towards Viksit Bharat 2047, highlighting the role of data in accelerating this goal. Ministries across the central government and states were also inspired to undertake data-based initiatives for national development. Several exploratory projects were underway, focusing on enhancing the use of data and evidence in the policymaking process by the Indian government. These initiatives span various ministries, departments, and sectors, aiming to integrate data-driven insights into decision-making to improve governance outcomes. However, many of these projects remain outside public knowledge, either due to their experimental nature, institutional sensitivities, or the absence of formal publication mechanisms. As a result, while the momentum for evidence-based policymaking is growing, the lack of transparency and public access to these efforts poses challenges for broader collaboration, replication, and accountability. The scope of our work expanded – from India's SDG journey, we could map the objectives towards Viksit Bharat.

Amitabh Kant, former CEO of NITI Aayog, led the initiative to come up with the first version of best practices across states, named "State Forward". Through this compendium, NITI Aayog created a repository on good governance and best practices in sustainable and equitable development across all states. This repository served a dual purpose: to appreciate the efforts of the officials involved in these initiatives and to serve as a good inspiration for other states to adopt those best practices. The editors of this volume were inspired by this initiative and envisioned a compendium on the use of data science and data-based methods that aided in policy making and governance. The numerous interactions on the use of data (including alternative big data) with Dr. Arvind Panagariya and Dr. Rajiv Kumar, both former vice chairmen of NITI Aayog, helped in gaining a deeper appreciation of these topics and the need for this edited volume to accelerate their adoption in India.

This edited volume comprises contributions from multiple authors, each offering distinct perspectives and insights on the subject matter. While no amount of gratitude by the editors towards the contributors can be enough, we thought it may be appropriate to dedicate the book to the army of data scientists in India who will help us achieve our SDGs.

The editors wish to express their sincere appreciation to Sonia Chauhan for her meticulous proofreading and editorial support. Her efforts in harmonizing the structure and formatting of the chapters have significantly enhanced the coherence and readability of the volume, thereby ensuring a seamless experience for the reader.

Avik would like to express his sincere gratitude to the Indian School of Business (ISB). It was there that his active involvement in capacity-building initiatives, particularly in teaching senior government officials and development sector professionals the application of data-driven methodologies for policy analysis and decision-making, shaped many of the ideas. This engagement not only deepened his understanding of the practical challenges and opportunities in applying data-based approaches within governance frameworks but also significantly informed the editorial process of this volume. The experience contributed to refining the chapters, focusing on methodological aspects and relevance, thereby reinforcing the adoption of evidence-based practices throughout the compilation. The editors would like to convey their sincere appreciation to the professors from the ISB Institute of Data Science (IIDS) and the Bharti Institute of Public Policy (BIPP) for their feedback on the chapters and narrative of this edited volume.

Bappa is immensely grateful to his colleagues and the team at Great Lakes Institute of Management, Gurgaon (GLIMG). The Centre of Excellence for Sustainable Development has supported this book in many ways, financially and otherwise. The role of students at GLIMG, as well as the University of Ulm, Germany, has been significant, as they have often been active discussants on some of the cases presented in this collection.

Finally, we would like to appreciate all the efforts put in by Rajesh Dey, editor, T&F. Without his coaxing and scolding, India would have achieved its goals much earlier than the book getting published. We hope the readers enjoy the compilation and, most importantly, the data science community gets inspired to help India achieve her goals.

Editor Bios

Avik Sarkar comes with over 20 years of experience across industry, government, and academia and is currently associated with the Indian School of Business, working and teaching in data, emerging technology, and public policy. He served as Officer on Special Duty (OSD) at the National Institution for Transforming India (NITI Aayog), where he headed the Data Analytics Cell, which was involved in envisioning the role of data and analytical methods in the policy-making process in India. At NITI Aayog, Dr. Sarkar helped in developing India's first AI strategy and roadmap for the use of data, analytics, and artificial intelligence for governance and policymaking across various sectors for India's inclusive growth and led efforts towards setting up the first high-performance computing-based data analytics lab and energy modeling unit at NITI Aayog.

His prior industry experience includes working with organizations such as Accenture, IBM, Nokia, NASA, Persistent Systems, and Zycus. At Accenture Consulting in Singapore, he contributed to various data and analytics-related engagements with the Singapore government, focusing on efficient reporting and resolution of municipal issues, evaluation and monitoring of secondary school education, and improving operational efficiency in immigration and customs. While at IBM, Dr. Sarkar contributed significantly to the development of the Monte Carlo simulation for SPSS and the predictive maintenance and quality solution for the manufacturing sector. Dr. Sarkar holds a PhD from the Open University, UK, a master's from the Indian Institute of Technology (IIT) Bombay, and a bachelor's from Calcutta University. Dr Sarkar has authored several technical publications and technology patents.

Bappaditya Mukhopadhyay (PhD, Economics, Indian Statistical Institute, New Delhi) is currently Professor of Analytics at Great Lakes Institute of Management and Visiting Professor, University of Ulm, Germany. His area of interest is public policy and development economics. As part of the Centre of Excellence for Sustainable Development at GLIMG, he has developed various indices pertaining to sustainability. Some of his notable public policy interventions included estimating the effect of microfinance, creating a white paper for transitioning to a cashless journey, designing direct benefit transfers, and linking alternative livelihoods in the face of climate risk, among others.

Contributors

S Anand
CEO, Gramener
Hyderabad, Telangana, India

Ayushi
Strategy and Innovation Consultant
Digital Transformation
Germany

Anjali Bansal
NITI Aayog
New Delhi, Delhi, India

Juzer Bharmal
Lookman Electroplast Industries Limited
Maharashtra, India

Anindya Bhattacharya
Executive Director
The Celestial Earth
New Delhi, Delhi, India

Bishakha Bhattacharya
Head of Public Policy
Wipro and Former Head of
Government Relations, IBM
New Delhi, India

Rwitwika Bhattacharya
Swaniti Initiative
New Delhi, Delhi, India

Debmalya Roy Chowdhury
World Wildlife Fund (WWF), India
New Delhi, Delhi, India

Poornima Dore
Director
Analytics Insights & Impact Tata Trusts
Mumbai, Maharashtra, India

Ashish Gautam
Cofounder Farmbox and Former Associate
Director, IBM
New Delhi, Delhi, India

Saloni Goel
Senior Director ESG, Coca-Cola, India
Former Head, Circular Economy, UNDP
New Delhi, Delhi, India

Raj Kumar Gupta
Manager (Retd) Uttar Pradesh
Development Systems Corporation Ltd. (UPDESCO)
Lucknow, Uttar Pradesh, India

Anand Kulkarni
Former Consultant
Government of Maharashtra
Mumbai, Maharashtra, India

Aswin Kumar
IT Project Manager
Vassarlabs IT Solution PVT. LTD.
Hyderabad, Telangana, India

Krishna Kumar
CEO, CropIn
Bengaluru, Karnataka, India

Nikhilesh Kumar
Co-founder & CEO APAC Region
Vassarlabs IT Solution PVT. LTD.
Hyderabad, Telangana, India

Chandrakiran Mallarapu
Co-founder
Vassarlabs IT Solution PVT. LTD.
Hyderabad, Telangana, India

Paresh Manohar
Program Manager
Data Driven Governance, Tata Trusts
Mumbai, Maharashtra, India

Saptarshi Mondal
Geospatial Product Development Manager
Vassarlabs IT Solution PVT. LTD.
Hyderabad, Telangana, India

Prerna Mukharya
CEO, Outline India
New Delhi, Delhi, India

Bappaditya Mukhopadhyay
Great Lakes Institute of Management
Gurgaon, Haryana, India

Harsh Nisar
Data Scientist Ministry of
Rural Development
New Delhi, Delhi, India

Praveen Pankajakshan
Chief AI Scientist
UrbanKisan and Former
Chief Scientist, CropIn
Bengaluru, Karnataka, India

Jayavaradhan Sambedu
CEO, Curl Analytics
Hyderabad, Telangana, India

Prukalpa Sankar
Founder Atlan and Former Cofounder
SocialCops
San Francisco, California, United States

Dipyaman Sanyal
CEO, dono Consulting
New Delhi, Delhi, India

Raunak Saraf
OSD, Government of Maharashtra
Mumbai, Maharashtra, India

Avik Sarkar
Indian School of Business
Hyderabad, Telangana, India

Sampurna Sarkar
World Wildlife Fund (WWF), India
New Delhi, Delhi, India

Aawanti Singh
Swaniti Initiative
New Delhi, Delhi, India

Aparajita Singh
Swaniti Initiative
New Delhi, Delhi, India

Kamaljeet Singh
Head of Cloud Business
JioBusiness and Former Practice Head, SAS
Mumbai, Maharashtra, India

Aakash Solanki
Joint Secretary
Ministry of Rural Development
New Delhi, Delhi, India

Gayatri Srinivasan
Lookman Electroplast Industries Limited
Chennai, Tamil Nadu, India

Sugata Srinivasaraju
Journalist-Author-Columnist
Bengaluru, Karnataka, India

Simi Thambi
Climate Economist
FAIRR Initiative and Former Specialist
Energy and Climate Change, NITI Aayog
London, United Kingdom

Akhilesh Tilotia
Co-Founder Thurro and Former OSD
Ministry of Civil Aviation
Mumbai, Maharashtra, India

Alka Upadhyaya
Researcher
Ministry of Rural Development
New Delhi, Delhi, India

Richa Verma
Principal Product Manager
AuraSell AI and Former Resident Entrepreneur,
SocialCops
San Francisco, California, United States

Shruti Yerramilli
Research Associate
Data Driven Governance, Tata Trusts
Mumbai, Maharashtra, India

Introduction

Bappaditya Mukhopadhyay and Avik Sarkar

On August 15, 2022, while addressing the nation on her 75th Independence Day, Prime Minister Narendra Modi spoke about entering the Amrit Kaal (the 25-year period leading to 2047) and called upon citizens to work collectively toward building a Viksit Bharat (Developed India) by the time the nation celebrates its 100th year of independence. Viksit Bharat is an aspirational vision that seeks to transform India through an inclusive, self-reliant, innovative, and sustainable India.

However, as India aims towards being Viksit Bharat by 2047, we must clearly understand the roles each players need to play towards achieving it. For an emerging economy, these in turn define clear roles the citizens, the industry, the government, the judiciary, and the civil society must play.

Our progress towards Viksit Bharat should have well-set targets, with timelines and regular scrutiny of where we were, where we are, and the road ahead. The first challenge is to define the metric of Viksit Bharat. Luckily, we have an all-encompassing framework. Improved scores in the Sustainable Development Goal (SDG) framework are the right way of getting there. For example, mere increase of per capita income cannot be the yardstick unless accompanied by low poverty, low inequality, and protection of various rights of citizens and natural resources. To achieve the goals, both the public and the private have important roles. The private sector can play its role where the market is able to provide the solutions, while the government must set up the proper structure (enable markets) and step in whenever there is market failure. One all-encompassing strategy that the government is almost always provided to use is providing social goods. Providing social goods has in the past been challenging, largely owing to a lack of information on the true benefits, the beneficiaries, and efficient delivery. As we shall see across the chapters, various initiatives that are driven by the government on its own or with private partnerships yield better outcomes when combined with data science.

0.1 DATA SCIENCE

Data science has gained prominence in recent years due to the explosion of digital data and the development of powerful computational tools and algorithms. The advent of big data and advancements in machine learning and artificial intelligence propelled data science to the forefront of various industries. Today, it plays a pivotal role in shaping decision-making processes, driving innovation, and solving complex problems across sectors like finance, healthcare, e-commerce, and more, reaffirming its significance in our data-rich world.

By analyzing and interpreting data, data science enables entities to uncover valuable insights, optimize processes, and enhance efficiency, which helps them make decisions that were not previously made. How does that happen? Data science encompasses a wide range of techniques and methods for extracting insights and knowledge from data. Data science combines engineering to extract various types of data, uses statistical techniques to analyze them, and thereafter uses specific domain knowledge to draw insights. The journey of data science starts with data collection, which involves gathering data from various sources, such as databases, web scraping, sensors, surveys, and more. Data analysts and data scientists must ensure data quality, consistency, and relevance. As the next step, data cleaning and preprocessing, involves handling missing values, outliers, and inconsistencies, thus preparing the data for initial analysis, termed exploratory data analysis (EDA).

Exploratory data analysis involves visually and statistically exploring data to understand its characteristics, identify patterns, and gain insights into potential relationships and trends. EDA can take

DOI: 10.1201/9781003487531-1

various names in the data science world, ranging from descriptive analytics to summary statistics, descriptive statistics, reporting, dashboards, and data visualization. After EDA, data science starts analyzing the data, which is used to uncover relationships within the data as well as to make specific predictions and recommendations either to arrive at an outcome or to avoid it.

Specifically, data science can generate new insights in the most critical social sectors that cater to the overall development of humankind, such as healthcare, education, poverty alleviation, climate change, governing systems, providing security, and so on. Data science primarily depends on large pools of data which are collected through various transactions or administrative systems or data collection through government machinery.

Today, every organization – private or public – is a repository of massive amounts of data. This forms a fertile ground for data scientists to analyze the data with these organizations and generate critical insights that improve systems so that the quality of life is improved for the largest number of people in the largest possible way. For example, let's take a government entity or an NGO providing food assistance to homeless and marginalized communities. Data science helps us find out how many people may need assistance each month based on the data collected. Were these problems not present earlier? Were they not being addressed? So why do we need data science now?

The answer lies in the scope of data science. For example, to identify those who were below an economic threshold earlier, the only way to address that was through collecting extensive data, doing laborious, expensive, time-consuming, and often corruptible surveys. Policies based on such data would often involve "leakages". Genuine beneficiaries may often get crowded out by others. Data science may involve simple metrics, analyzing mobile usage to address the economic status fairly well. Such exercises often result in significant improvement in policy outcomes. Data science can analyze the data collected by various projects to bring out the impact of each program. Once we know the impact, we can further drill down and come up with possibilities of why a certain project succeeded or failed in a particular geographical territory. This can furthermore lead to establishing better-suited projects and programs that help communities in the best possible ways and use the resources more effectively. While most usage of data science is for business decisions, of late, governments across the world are using it to make better decisions. It is being extensively used to design and deliver social goods.

At its core, a social good is an initiative that attempts to help or benefit the greatest number of people in the largest way possible. Today, we can no longer think of achieving social goods without making them sustainable for the future generations of humans who will walk the planet after us. The road map towards achieving Viksit Bharat would therefore align with the SDGs. Since social goods and Sustainable Development Goals have the same goal, it is logical that they go hand in hand.

Overall, the UN describes 17 SDGs that address five pillars – people, prosperity, planet, peace, and partnership. Zooming in on each pillar, we see the main SDGs tackling global issues that plague humanity, such as poverty, hunger, and inequality, and aiming towards building better systems that align with good health, quality education, economic growth, innovation, sustainable communities, and institutions that promote peace and justice. Further, there is a vital concern that the SDGs want systems to work towards – climate preservation and protection. For this purpose, specific goals include responsible consumption and production, clean water and sanitation, climate action, clean energy, and life on land and water.

This book carefully collates some case studies that were taken by the government (both at the state level and the center) to deliver such social goods either on their own or in partnership with private entities. These initiatives play an important role in India's move towards the SDGs.

0.2 SOCIAL GOOD AND SUSTAINABLE DEVELOPMENT GOALS

In September 2015, 193 countries came together at the United Nations to adopt and commit to 17 Sustainable Development Goals with 169 sub-goals.[1] India, a signatory to the 2030 Agenda for

FIGURE 0.1 The 17 Sustainable Development Goals (source: UN website).

Sustainable Development, has launched several developmental programs to meet all 17 SDGs in a time-bound manner. As the most populous nation in the world and the world's fifth-largest economy, India's success in achieving the SDGs will largely determine global outcomes.

The book chapters are divided across five sections. The sections are carefully designed to cover broad contours of policy and governance, as well as specific issues. The section details are briefly presented in the following.

0.3 PLANNING AND GOVERNANCE

Planning and governance refer to the processes and mechanisms by which a government organization or body sets strategic objectives, develops policies, and manages resources to achieve the well-being and development of its citizens. It is a forward-looking process identifying the actions, resources, and timelines necessary to reach desired outcomes. Planning can occur at various strategic, tactical, and operational levels. In the context of government, it often refers to urban, economic, and social planning, among others. Effective planning is essential for allocating resources efficiently, responding to challenges, and ensuring sustainable development. This can be planning for the entire country or a specific state, district, or village in terms of strategic, urban, economic, and social, and other areas.

The section on planning and governance touches upon the various Sustainable Development Goals across the various chapters:

- Goal 1: No poverty
- Goal 2: Zero hunger (no hunger)
- Goal 3: Good health and well-being
- Goal 10: Reduced inequalities
- Goal 16: Peace, justice, and strong institutions

This section contains chapters that present how data is used to create a framework to track and evaluate India's SDG readiness. In India, an effort to incorporate data-driven governance maps the broad country-level goals down to the local (Panchayat) decision-making level. Section I highlights some interventions that present India's commitment to Sustainable Development Goals at village-level planning.

In Chapter 1, Anand Kulkarni and Raunak Saraf show how data-driven decision-making helps optimize the usage of Chief Minister's Relief Funds (CMRF) in Maharashtra. CMRFs are designed to provide relief to the state's citizens in distress by providing financial assistance to people affected by floods, droughts, accidents, and natural disasters and, at times, during disease outbreaks. Shortage of funds means several requests are not fulfilled, and excess funds would require further justification from the donors. The CMRF Dashboard in Maharashtra has provided a solution to all these issues in the form of a forecast data analysis model. The forecast data analysis model calculates the existing and required funds and understands the utilization and provision of funds for social welfare in a simple, clean, and navigable manner. Thus, the CM Dashboard helps officers optimally plan funds disbursement across various factors such as time, place, and incidents.

The next chapter 2 in this section is by Anjali Bansal. The chapter documents India's journey towards developing the SDG India tracking index. Driven by the Ministry of Statistics and Programme Implementation (MoSPI), the National Statistical Unit, and NITI Aayog, the government's premier policy think tank, the SDG India Index tracks the states' progress on 16 SDG parameters like health, education, economic growth, climate change, and environment. The chapter documents the process of transforming the national-level SDG indicators into state-level indicators that would appropriately measure the SDG, followed by the challenges involved in acquiring recent data for the purpose of creating the index and finally discussing the process involved in creating the first State SDG Index for India.

A couple of chapters in this section present case studies pertaining to village-level interventions using data science. Rwitwika Bhattacharya and her team carried out a massive data collection exercise across schools, houses, panchayats, and health centers in 49 gram panchayats of Balasore, Orissa, to create development plans for each village (Chapter 4). This helped in targeted development through community governance. Based on the data, a priority matrix was created across sectors, based on which village development plans were framed. For example, data revealed that 18% of villagers did not hold an Aadhar Card, and only 8.2% of households were using liquefied petroleum gas (LPG) as the primary cooking energy source. Village officials then carried out targeted interventions to alleviate the problems of the villagers, and the previously underused gram panchayat funds were put to these data-backed, actionable tasks. This resulted in almost the entire population being covered by Aadhaar, with substantial LPG coverage. In another chapter, Poornima Dore and the team at Tata Trusts present how data was used to help the intended beneficiaries demand accountability from institutions in charge of program delivery at Chandrapur village in the Nagpur division of Maharashtra. Data, evaluation, learning, technology, and analysis (DELTA) was a framework employed for a participatory approach in microplanning so that the development plans truly reflect the needs of the beneficiary community (Chapter 3). The framework encompasses collecting and analyzing datasets for institutions (schools, anganwadi, village, gram panchayat, health centers), households, and individuals. The DELTA Pilot in Chandrapur (Maharashtra) was conducted in three of its blocks, including 115 gram panchayats. Subsequently, an intensification program was initiated towards the implementation of the village development plan, which was prepared based on the primary survey, with the intent to encourage people to analyze the current situation of their village development and seek accountability from elected representatives and government officials. The DELTA app allowed all datasets to be put together on one dashboard – which means, for example, we can not only see how many individuals reported a case of malaria in the recall period (through survey data), but we can also see how many such cases were recorded at the Primary Health Center (institutional information) and may be able to recognize one of the main hurdles in early treatment like non-availability of ASHA workers in the concerned village/hamlet (information obtained through Participatory Rural Appraisal exercises).

The final two chapters in this section relate to two essential components of democracy and governance – elections and the Assembly proceedings. In Chapter 5, Sugata Srinivasaraju shows how big data was analyzed and sifted to present an interactive experience to readers during the Karnataka elections of 2013. The focus was on governance and the government's welfare programs. To make the information attractive to voters, all data relating to candidates, including their wealth and qualifications, was collected and then presented in the form of colorful roulettes, cartograms, treemap visualizations, and bubble charts that provided critical information to the reader in an entertaining and insightful manner. Such dissemination of candidate-level information prior to elections is now an important template before most elections. In Chapter 6, Anand Kulkarni and Raj Kumar Gupta document how, using the Intelligent Search Mobile Application, a plethora of documents related to Uttar Pradesh Legislative Assembly proceedings could be retrieved and analyzed. This considerably eases research and analysis of discussions in the assembly by saving time on manual or physical searches. Previously, these searches were conducted in libraries to retrieve information by authorized personnel, taking a humongous time for the entire process. The application was developed based on scanning, annotating, and indexing all the digitized assembly proceeding books, videos, newspaper clippings, and gazettes. This mobile app enables members of the Legislative Assembly (MLA) and staff to access the documents of the Vidhan Sabha proceedings at their fingertips, making it efficient to find past information and perform official duties.

The previous chapters indicate that India is successfully engaging data science in all aspects of its democracy and governance, from elections and their outcomes to commitment on the global stage, down to village-level program implementation and participatory democracy and governance.

The next section of the book covers human capital. For India to progress, it is vital that we invest and manage our human capital well. The chapters highlight some data-driven initiatives towards that.

0.4 HUMAN CAPITAL

> It is critical to prepare the "demographic dividend" of the country as per the demands of the changing job roles.
>
> **– Prime Minister Narendra Modi, February 21, 2022**

Since coming to power, Prime Minister Narendra Modi has often invoked the phrase "demographic dividend" in his speeches and stressed the need to harness it for the nation's economic growth and youth empowerment. Demographic dividend occurs when the ratio of the working-age population is high and the dependency ratio in terms of the proportion of children and elderly people is low. This is expected to have a positive influence on the growth and development of society and the country, something that is critical for India's growth story and future. India entered the demographic dividend opportunity window in 2005–06 and will remain there until 2055–56. This period is expected to see a working age ratio exceeding 150% and dependency ratio lower than 66.7%. Working age ratio measures the proportion of dependents (those under 15 or over 64) to the working-age population (ages 15–64). Further, the highest working age ratio for India will be between 2021 and 2041, with the peak in the working age ratio occurring in 2031[2] – more precisely, India is said to have entered into 37 years of demographic dividend, lasting from 2018 to 2055, due to a declining number of dependents compared to the working population.[3] However, to harness the demographic dividend, it is crucial that India invest and manage the most important factor: human capital. Investing and managing human capital involves having a healthy population, a population that is educated, and education that maps to critical skill development.

Human capital refers to individuals' knowledge, skills, experience, and abilities, which can be viewed as an asset or resource. This concept recognizes that people can contribute to economic and societal growth through education, training, and life experiences. A healthy individual is often more productive and has a higher capacity for learning and innovation. Formal education, including degrees and certifications, is fundamental to human capital. It equips individuals with knowledge and problem-solving skills. Skills acquired through training and on-the-job experience contribute to an individual's human capital. These can range from technical skills to soft skills like communication and leadership. The knowledge and expertise gained through work experience are integral to an individual's human capital. Accumulated experience can increase one's value in the job market. The SDGs related to human capital covered in this section of the book are listed as:

- Goal 3: Good health and well-being
- Goal 4: Quality education
- Goal 8: Decent work and economic growth
- Goal 9: Industry, innovation, and infrastructure
- Goal 16: Peace, justice, and strong institutions

The chapters in this section present case studies that use data science to address these issues at the national and state levels. While India's health expenditure has been increasing steadily (current health expenditure per capita, in Purchasing Power Parity, PPP [current international $] 86 USD in 2000 to 211 USD in 2019), as a percentage of GDP, the share is falling (from 4% to 3% during the same period). During this period, domestic general government health expenditure per capita, PPP (current international $) has risen from 18 USD to 69 USD. While the growth in absolute terms looks impressive, there are worrying signs. The proportion of the population spending more than 10% of household consumption or income on out-of-pocket health care expenditure (%) rose from 15% to 17% during this period, while the proportion of the population spending more than 25% of household consumption or income on out-of-pocket health care expenditure (%) increased from 3% to 6%. While an increasing out-of-pocket health expenditure suggests that a greater share of GDP must be allotted to health, India has made some important progress over the last two decades. This partly explains why the SDG2 gap between the world and India is reducing. Some key statistics are as follows: Maternal mortality ratio and prenatal as well as under-5 mortality rates have reduced dramatically from 370 to 145, 45 to 24, and 92 to 39, respectively, between 2000 and 2017. A large part of this success is owing to the fact that births attended by skilled health staff as a percentage of total births nearly doubled (from 42% to 82%) between 2000 and 2017. Immunization against diphtheria, pertussis, and tetanus (DPT), hepatitis B, and measles in children has also increased significantly (from 50% of children to about 90% now). India has made some progress in the areas of early childhood development. The prevalence of severe wasting, stunting, and underweight (% of children under 5) decreased steadily from 48, 20, and 43 in 2006 to 35, 14, and 33%, respectively, in 2017. Among the adult population, mortality from CVD, cancer, diabetes, or CRD between exact ages 30 and 70 has dropped from 25% to 22%. Further, the incidence of malaria (per 1,000 population at risk) and tuberculosis (per 100,000 people) reduced from 19 to 3 and 289 to 189 between 2000 and 2020. Some of India's recent successes in health outcomes, especially in the areas of women's and children's health attainment, have been driven by data interventions.

In this section, Raunak Saraf and Anand Kulkarni present a case study on the Child Health initiative in Maharashtra (Chapter 9). The Department of Women and Child Development, Government of Maharashtra, focuses on promoting the survival, protection, development, welfare, and participation of women and children in society and the economy in a holistic manner. Earlier, the information was collected manually by enumerators. The usual errors involving manual data collection, coupled with moral hazard problems, made the data unreliable. The indicators were highlighted based on such collected information. The method was time consuming. The efforts were manual; hence the accuracy was also not remarkable in terms of KPI values. In the case of child health cluster analysis, the dashboard of the Women and Child Development Department gives a comprehensive view of the analyzed data. The integrated data in terms of regions, disabilities, duration, and other parameters is now presented in the form of statistical graphs. This graphical representation enables the officers to manage the data and obtain clear results. These results help in better planning and execution of the related schemes. Thus, the Chief Minister Dashboard, Maharashtra, is designed to give officials health and disability indicators of women and children. The dashboard helped in monitoring the KPIs defined for each department. Comprehensive graphical reports at daily as well as monthly frequency generate insights from the dashboard to help the government plan better and more targeted schemes and policies that combat these specific disabilities in specific districts.

In Chapter 8, Kamaljeet Singh presents the usage of data in preventing the outbreak of the coronavirus in Odisha. The government of Odisha collected data from different healthcare departments, labs, hospitals, outpatient departments (OPDs), inpatient departments (IPDs), staff members, and ICMR forms. This data was then assessed and reviewed to come up with evidence-based planning, monitoring, and supervision of the state population's health and a real-time view of COVID spread in the state. The system offered insight into disease outbreaks, prediction, and prevention within Orissa and was adopted by various other state governments.

The status of education and skill attainment in India has also increased manifold during the last two decades. The literacy rate among the population over 15 years has increased from 60% to 75% and primary completion rates from 71% to 96%, with similar trends visible for secondary and upper secondary enrolments, too. However, the progress is not as significant when it comes to pupil–teacher ratios. The ratios have been between 24 and 34 at almost all levels, with minor improvements. However, the more worrying trend is that the total share of youth not in education, employment, or training (% of youth population) is at 27%, a marginal improvement from 31% over the last two decades. The aspirations of an emerging economy and encashing the demographic dividend would require a larger skilled labor force.

The next three chapters in this section highlight the role of data science in education and skill development. In Chapter 10, S. Anand demonstrates how the appropriate use of data-driven decision-making improves learning outcomes for students in school. The National Achievement Survey conducted by the National Council of Educational Research and Training (NCERT) surveys over 20 lakh students across standard 3rd, 5th and 8th in over 1 lakh government schools across every state. Findings from the survey helped in finding the critical behavioral tendencies of students that can improve their academic performance. A student questionnaire of 60 questions, divided into four categories: demographic, social, behavioral, and educational preference, was administered, and the results gave interesting insights. For example, it was found that children who play games perform better at mathematics, and children who have access to books at home and moderate watching of television improve academic performance. Children who do not watch any television, surprisingly, perform the worst. Taking private tuition has little to no impact. Findings like this give important micro-level interventions. For example, it turns out that for those below the poverty line, computer usage is more important than reading books, so investment in computers for schools serving those below poverty levels is a more critical initiative so that these children get access to computers from an early age.

In Chapter 11, Dipyaman Sanyal looks at data-driven initiatives for skill development across India by the National Skill Development Corporation (NSDC). This analysis helps fulfill NSDC's primary objective of supporting the skilling system and upgrading skills to international standards, using a data analytics pyramid. A large amount of data is fed into the data analytics pyramid right from the bottom level, that is, the status of skills, and then moves upwards to diagnose, predict, and then prescribe upskilling principles. The dashboard offered visual analytics that were utilized towards predictive and potentially prescriptive analytics. All data were completely anonymized and then processed to generate valuable data insights such as placement rates across age groups, states, sectors, and industries. Key insights were generated around formal and informal placement rates, course length, and the technical education of candidates.

Finally, in Chapter 7, Harsh Nisar, Aakash Solanki, and Alka Upadhyaya identify an interesting bias that often plagues policy making and allocation of funds disproportionately towards urban areas. The Ministry of Rural Development set up a flagship scheme, the Deen Dayal Upadhyay Gramin Kaushalya Yojna (DDUGKY), specifically to impart technical skills for employment to the rural population, for youth in the age group 15–29 years. This scheme operates in a public–private partnership model across various states to achieve India's skilling goals. However, data analysis showed that the greater the proportion of urban population a district had, the more money allocated for the rural skilling program. This meant a certain urban bias. Based on the analysis, an index was created using the Socio-Economic Caste Census 2011 and Census 2011. This index ensured that a district is rewarded if it catered to a higher number of deprived households and used negative reinforcement for a district based on urbanization. This index formed the basis for further allocation of youth for the rural skilling program. The tool allowed the state to enter district-wise targets for each Project Implementation Agency (PIA) in a matrix, and it would instantly update and highlight the total deficit/overflow in each district. The state effectively used the tool and allocated targets to skill more than 20,000 rural poor in a more equitable manner than previously done. As a success story, this was demonstrated nationally to all the states, and similar biases were found nationally. States were supported by the Ministry of Rural Development to ratify/balance their targets, and a policy under DDUGKY for the purposeful allocation of targets across districts was drafted.

This section highlights the possible data science–driven interventions towards improving aspects of human capital in terms of granular analysis of child health issues or managing operations during the COVID-19 in Chapter 8 by Kamaljeet Singh. Data science shows the possibility of understanding trends impacting student learning outcomes or allocating funds for skilling programs. This section will be an inspiration to apply data-based methods in various aspects related to healthcare, education, and skilling.

For us, for a long time, agriculture will remain the backbone of the economy that will still absorb a substantial proportion of employment. However, there is a tremendous imbalance between the GDP contribution of the sector (less than 20%) ands its significant contribution in employment (exceeding 40%). The goal eventually will be to reduce its dependence and vulnerability. The main intervention of data science here is easing climate vulnerability in the sector. The next section has a collection of chapters addressing that.

0.5 AGRICULTURE, CLIMATE AND ENVIRONMENT

From a government perspective, agriculture is a critical sector that encompasses policies, programs, and activities related to the cultivation of crops, raising livestock, and the overall management of rural and agricultural activities. India's economy has become more diverse and industrialized; agriculture remains a foundational sector supporting millions of people's livelihoods and contributing to the nation's overall well-being. A large part of agriculture in India is susceptible to climate change.

Variability in rainfall can impact crop yields and food production. Climate-related events like droughts or floods can significantly impact food production and supply, potentially leading to food security issues. India is prone to extreme weather events such as cyclones, floods, and heatwaves. Changes in climate patterns can lead to more frequent and severe such events, causing extensive damage. India's major rivers, including the Ganges and Brahmaputra, are fed by the Himalayan glaciers and seasonal monsoons. Any climate change affects water availability for agriculture, drinking, and industrial use. Climate change also impacts ecosystems, wildlife, and biodiversity, which are critical to the agricultural sector. Shifts in temperature and precipitation patterns can disrupt ecosystems and threaten species. The SDGs related to agriculture and climate covered in this section of the book are:

- Goal 2: Zero hunger (no hunger)
- Goal 13: Climate action
- Goal 15: Life on land

India ranks second worldwide in farm outputs. As of 2018, agriculture employed more than 50% of the Indian workforce and contributed 17–18% to the country's GDP. In 2016, agriculture and allied sectors accounted for 17.5% of the GDP (with about 41.49% of the workforce). India also ranks first in the world with the highest net cropped area, followed by the US and China. The total agricultural commodities export was US$3.50 billion in March–June 2020. While Indian agriculture has performed fairly well during the last two decades in terms of record production and doubling its share in global trade in agriculture, the sector is currently saddled with issues, especially related to sustainability, adoption of new agricultural technologies, and, perhaps most importantly, income levels of the population dependent on farming, which require long-term reorientation of food, agriculture, and farm policies. Some of the areas where big data analytics provides innovative solutions relate to weather forecasting and crop monitoring, as well as predicting possible crop failures and ways to address them. With climatic factors continuing to have a significant impact on agricultural productivity in India, rising temperature, along with increased occurrences of extreme weather conditions, has made climate change a major threat to Indian agriculture and productivity loss. Long-term co-movements show that the change in different climatic variables has different implications for rabi and kharif crops. Overall, the preliminary findings indicate that maximum temperature impact is more prominent during the kharif season, while minimum temperature anomalies impact rabi crop yields more.

In the last two decades, significant progress has been made towards using AI–ML to improve agricultural outcomes. How do big data and data science aid decision-making in agriculture? These techniques use unstructured data like satellite imagery, drones, sensor data captured through IOT devices, combined with traditional data related to the soil quality, fertilizer input, seed types, etc. and The broad areas where big data is used are crop monitoring and forecasting, water management and harvesting, soil analysis, and weather forecasting. By analyzing historical data about weather patterns and crop yields, models are developed that can predict how a particular crop will fare in the future under different conditions, helping farmers make decisions about what crops to plant and when to harvest them. Further, farmers can now use sensors and drones to collect data about their crops, which can be analyzed to identify problems early on and take corrective action to improve yields and avoid losses due to pests or diseases. Data science is also increasingly being used for water management and soil analysis. The former is done by collecting data about weather patterns, soil moisture levels, and irrigation systems, while sensors are used to monitor soil quality.

In Chapter 16, Jayavardhan Samvedu demonstrates how data analytics is used for both the short-term and long-term forecast. Short-term forecasting has higher accuracy due to the low probability of sudden changes in the system. For example, the probability of rain in the next hour is very low if it is not raining right now. However, forecasting whether it will rain in the next hour needs many dynamic parameters, not only from land weather stations but also from satellites and balloons. The dynamics of the atmosphere are complex, and a heavy numerical model is required, which uses differential equations and runs on supercomputers. In contrast, long-term forecasts are less accurate and less dynamic in comparison to short-term forecasts. They require less data, different variables from all over the world, and far less complicated computation. In the initial stage, the model was built to forecast rainfall for Brazil because the US market is heavily dependent on the agricultural produce of Brazil, and hence readily available validation exercises can be used to test the model. For example, Brazil exports a major chunk of the world's coffee, and many markets depend upon the amount of coffee imported from Brazil. However, by using variables precisely according to the dynamics of the atmosphere in the Indian region, the model performed better than the Brazilian model. Accuracy for India is more than 90% for both 1-month and 3-month forecasts.

In Chapter 15, Ashish Gautam and Bishakha Bhattacharya document how the Centre for Spatial Analytics and Advanced GIS (C-SAG) and Tata Trusts jointly designed the concept of plots for a farmer-centric comprehensive agricultural advisory. They have developed a process to build, manage, analyze, and generate real-time GIS-based advisories at the individual plot level. This end-to-end process is being implemented in 532 villages of three districts in South Odisha. The process involves "disaggregating" a variety of spatial and non-spatial inputs – using complex data analytics methods at a plot level for a 360° evaluation of a farmer (owner/tiller of the plot). This, in turn, brings integrative spatial analytics support to the farmer. For meteorology analytics, C-SAG has been accessing forecast and measured data sets from the India Meteorological Department (IMD) and enhanced forecasts from IBM. C-SAG successfully monitored and alerted farmers about the Titli and Phethai cyclones on October 10, 2018, and December 17, 2018, respectively.

The next two chapters present a few case studies regarding how satellite imagery is being used in cropping patterns. In one chapter, Krishna Kumar (Chapter 13) and in the other Praveen Pankajakshan and Krishna Kumar (Chapter 14) present assessment methods and cases from studies across India. The Crop Cutting Experiment (CCE) is an efficient assessment method used to determine the expected yield in that agricultural cycle. Earlier methods consisted of randomly sampling areas where various soil and crop cycle factors were tested to indicate harvest criteria over a larger area. With the help of artificial intelligence and machine learning, digital solutions have been introduced in the CCE to help bring focused results. These digital solutions, SmartFarm and SmartRisk, use ground-level data and satellite imagery to identify plots that are appropriate for these experiments. A dedicated and highly skilled data science team analyses millions of data points and runs them through numerous criteria to zero in on farm plots that will provide the most accurate samples for the region. In this way, ambiguity is removed from plot sampling, making crop insights valuable and focused. There are three separate case studies documented in one of the chapters: the Jeevika World Bank Project, Mission 2020: Lakhpati Kisan – Smart Villages Project, and Seed Potato Traceability Project. These projects were launched across various states of India where technology was introduced in farming methods to increase livelihoods, provide farm advisory services, counter climate change, uphold the lives of tribal households, and improve seed quality. Farm data is captured at critical points to establish end-to-end traceability of the life cycle of crops starting from pre-harvest until final harvest. Raw data is processed, and farmers are provided with data insights, including weather alerts, pest forecasts, management practices, monitoring production and harvest, and adherence to export compliance. Using the technology, 80% of farm households implemented the advisories issued, 92% demonstrated climate-resilient agricultural practices, and 90% declared satisfaction with the satellite-based advisory.

In Chapter 12, Chandrakiran Mallarpu, Nikhilesh Kumar, Aswin Kumar and Saptarshi Mondal shows how an integrated digital solution, Andhra Pradesh Agriculture Information and Management System (APAIMS), was developed by the government of Andhra Pradesh to analyze various data sets and churn out meaningful advisories. The main stakeholder is the farmer, and all advisory and support services are created and offered keeping in mind the ultimate benefit of the farmer. The integrated platform captures information such as Soil Health Card data, localized weather data from automatic weather stations (AWSs), and soil moisture data from sensors. It superimposes all this information with satellite data-derived insights, information on the availability of water, geospatial data, crop phenology models, and so on. The APAIMS digital platform capabilities include assimilation of complex data backed by analytics and a knowledge engine to produce a holistic "farm-specific" advisory.

The final Chapter 17 in the section concerns SDG 15, that is, life on land. Debmalya Roy Chowdhury and Sampurna Sarkar document how camera traps were used to estimate the number of tigers in the Sundarbans Tiger Reserve in 2012. The idea was novel, as the habitat in the Sundarbans is different from other tiger reserves, and hence, one had to be innovative. Since 2006, tiger estimation exercises using camera traps are conducted every 4 years in India under the supervision of the National Tiger Conservation Authority (NTCA). However, in the first such exercise, the Sundarbans was left out, and the first all-India tiger status report published by NTCA in 2008 did not report tiger numbers from the Sundarbans. The second all-India tiger status report, published in 2011, reported 70 tigers in the Indian Sundarbans, although no camera traps were installed in the Sundarbans during the exercise. The estimate was based on radio collaring of some tigers in the mangrove habitat. After the number was published, the then–principal chief conservator of forest–wildlife (PCCF-WL) and chief wildlife warden (CWLW) of West Bengal expressed reservations regarding the estimated number of tigers and the methodology. The CWLW advocated for the development of a methodology appropriate for the Sundarbans.

0.6 ENERGY, WATER, AND SANITATION

The specific SDGs we address in this section are:

- Goal 3: Good health and well-being
- Goal 5: Gender equality
- Goal 6: Clean water and sanitation
- Goal 7: Affordable and clean energy

0.6.1 Energy

India is a major force in the global energy economy, with energy consumption more than doubling since 2000, propelled upwards by a growing population – soon to be the world's largest – and a period of rapid economic growth. India's continued industrialization and urbanization will make huge demands of its energy sector and its policy makers. Energy use on a per-capita basis is well under half the global average, and there are widespread differences in energy use and the quality of service across states and between rural and urban areas. The affordability and reliability of energy supply are key concerns for India's consumers.[4] On the production side, while oil production has hardly increased over the last three decades, coal production has increased steadily, keeping pace with energy consumption per capita.

India's growth in the economy has created a large affluent class, and thus, the generation and consumption of power have increased manifold. Post-1990, near-universal household access to electricity was achieved in 2019, meaning that over 900 million citizens have gained an electrical connection in less than two decades. There has been a rapid increase in both electricity generation as well as consumption. However, the growth of renewable energy lags behind the non-renewable production of electricity despite India's power sector being one of the most diversified in the world. Sources of power generation range from conventional sources such as coal, lignite, natural gas, oil, hydro, and nuclear power to viable non-conventional sources such as wind, solar, agricultural, and domestic waste. India was ranked fourth in wind power, fifth in solar power, and fourth in renewable power installed capacity as of 2020. India is the only country among the G20 nations that is on track to achieve the targets under the Paris Agreement.[5]

Data-driven decision-making aids both an integrated approach to energy planning and ensuring all households have access to energy. In chapter 18, Simi Thambi and Anindya Bhattacharya present data-driven integrated models that are used to display alternative futures with different fuel supply options. The chapter discusses the challenges and opportunities in using energy models effectively for policy making. Global best practices have been listed to demonstrate the status and importance of energy modelling across the world. Then the authors share insights gained from national and state-level energy modeling and inter-model comparison exercises to bring out their scope and some areas of opportunity. NITI Aayog has developed an in-house energy optimization model. In 2017, NITI Aayog took steps to develop an energy optimization model in-house with technical support from the International Institute of Applied System Analysis (IIASA) under the umbrella of the Technology Information Forecasting and Assessment Council's (TIFAC) membership in IIASA. The model built by NITI Aayog uses an optimization model based on a scenario assessment between baseline and Nationally Determined Contribution (NDC; presented before UNFCCC) scenarios, showing that by achieving its targets for increasing renewable capacity per NDC commitments, India can increase the share of renewables in power generation to 22% by 2032.

In Chapter 20, Prukalpa Shankar and Richa Verma show how various data sources were used to ensure all of India was covered by cooking gas distribution. The Pradhan Mantri Ujjwala Yojana

(PMUY), launched in 2016, aimed at providing LPG connections to BPL households in the country. Ten thousand new LPG distribution centers had to be opened in locations for almost every citizen in each of the 6.4 lakh villages to reach a distributor within 10–15 km of his/her residence. The chapter talks about the data science–based approach for deciding the ideal locations for the new distribution centers. The data of the currently operating LPG centers was put on a map along with other relevant criteria such as population, LPG penetration, and income levels. Then access was made the primary factor for deciding new center locations. The approach of the government was not based on market potential but opening centers in the vast blank of the nation's geography where people do not have access to LPG and thus must resort to other fuels for cooking. The authors use geospatial mapping along with clustering to solve the problem. This is a major intervention in improving gender-related outcomes, as the traditional cooking media (charcoal, kerosene, forest wood) had a major health impact on rural women.

0.6.2 Water

India is a water-rich country with 4% of the world's water resource. Rivers have been the heart and soul of India's growth and culture. Among them, 12 rivers are classified as major rivers, which cover about 253 mha of catchment area, and 46 are classified as medium rivers, with 24.6 mha of catchment area. Many of the river systems with their tributaries are perennial, and some of them are seasonal. A total of 3,237,021 km^2 of catchment area provides about 1869.35 bcm, out of which about 690.1 bcm is usable surface water.

The projected water requirement is 843 bcm in 2025, up from 710 bcm in 2010, out of which more than 70% is expected to be in irrigation and the remaining is likely to be shared among drinking (7%), industry (8%), energy (4%), and others (8%).[6] While in Section III, we address the issue of water management for irrigation, in this section, apart from agricultural usage, we will focus on the usage of data science in managing water for drinking and industry, too. As India is expected to grow at a rapid pace and will support the world's largest population soon, some problems lie ahead.

In Table 0.1, two important observations stand out from the last three decades of water usage data. One, a steady increase in the freshwater withdrawal (51.3% in 1990 to 66.5% in 2019) and an increased number of households having access to basic drinking water services (up from 80.5% in 2000 to 90% in 2019). With this trend, by the next decade, we may see the entire population having access to basic drinking water, which would mean a withdrawal rate exceeding 75% by that time.

Water availability per person is dependent on population of the country, and for India, per capita water availability in the country is reducing due to increase in population. The average annual per capita water availability in the years 2001 and 2011 was assessed as 1816 and 1545 cubic meters, respectively, which may further drop to 1486 and 1367 cubic meters in the years 2021 and 2031, respectively. Per the Ministry of Housing and Urban Affairs, 135 liters per capita per day (lpcd) has been suggested as the benchmark for urban water supply. For rural areas, a minimum service delivery of 55 lpcd has been fixed under Jal Jeevan Mission, which may be enhanced to a higher level by states.[7]

This requires managing water well. In Chapter 22 in this section, Chandrakiran Mallarapu and Nikhilesh Kumar show how the Andhra Pradesh Water Resources Information and Management System (APWRIMS) uses machine learning, artificial intelligence, and big data to depict real-time data on state water resources and present data-driven insights for decision makers of the state. Water is managed in the state through APWRIMS, which measures the total amount of water in the state and how it is spread across the area, including major and medium projects, the purposes for usage of water, groundwater and freshwater supply and depletion levels, water pollution levels, and strategies for water conservation in the state. While developing APWRIMS, during the data capture process, Vassar Labs collected information from approximately 1 million geo-tagged water bodies and 1.3 million geotagged bore wells, along with various layers like administrative, hydrological, soil, agriculture, satellite, aquifer, land use land cover (LULC), rivers and canals, drainage networks, digital

TABLE 0.1
Water Usage in India

Year	Annual Freshwater Withdrawals, Agriculture	Annual Freshwater Withdrawals, Domestic	Annual Freshwater Withdrawals, Industry	Freshwater Withdrawal as a Proportion of Available Freshwater Resources	People Using at Least Basic Drinking Water Services
1990	92.0	5.0	3.0	51.3	
1991	91.9	5.2	2.8	52.5	
1992	91.9	5.4	2.7	53.6	
1994	91.8	5.8	2.4	55.9	
1995	91.7	6.0	2.3	57.0	
1996	91.7	6.2	2.1	58.1	
1997	91.6	6.4	2.0	59.3	
1998	91.6	6.6	1.9	60.4	
1999	91.5	6.7	1.8	61.5	
2000	91.5	6.9	1.6	62.7	80.5
2001	91.4	6.9	1.7	64.2	81.0
2002	91.2	7.0	1.8	65.8	81.6
2003	91.1	7.0	1.8	67.3	82.1
2004	91.0	7.1	1.9	64.2	82.6
2005	90.9	7.1	2.0	64.6	83.1
2006	90.8	7.2	2.0	65.0	83.7
2007	90.7	7.2	2.1	65.3	84.2
2008	90.6	7.3	2.1	65.7	84.7
2009	90.5	7.3	2.2	66.1	85.2
2010	90.4	7.4	2.2	66.5	85.7
2011	90.4	7.4	2.2	66.5	86.2
2012	90.4	7.4	2.2	66.5	86.7
2013	90.4	7.4	2.2	66.5	87.2
2014	90.4	7.4	2.2	66.5	87.7
2015	90.4	7.4	2.2	66.5	88.1
2016	90.4	7.4	2.2	66.5	88.6
2017	90.4	7.4	2.2	66.5	89.1
2018	90.4	7.4	2.2	66.5	89.6
2019	90.4	7.4	2.2	66.5	90.0

Source: World Bank Development Indicators

elevation model data (DEM), and command areas. Approximately 3 million data points come every day from different sources like automatic weather stations (AWS), reservoir level sensors, groundwater level sensors, and river and canal level sensors to APWRIMS.

0.6.3 Sanitation

Despite being a rapidly growing emerging economy and witnessing gradual improvements in human development indices, India still faces the huge challenge of poor sanitation. In 1980, rural sanitation coverage was estimated at 1%. By 2018, it reached 95%. The share of Indians with access to improved sources of water increased significantly from 72% in 1990 to 88% in 2008.[9] However, many people still lack access to water and sewage infrastructure.

A few innovative approaches to improve water supply and sanitation were tested in India in the early 2000s. These included community-led total sanitation, demand-driven approaches in rural water supply, and public–private partnerships to improve the continuity of urban water supply. In 1999, a demand-driven and people-centered sanitation program was initiated under the name Total Sanitation Campaign (TSC), which has some similarities with community-led total sanitation (CLTS) but is not the same. It evolved from the limited achievements of the first structured program for rural sanitation in India, the Central Rural Sanitation Programme, which had minimal community participation. The main goal of the Total Sanitation Campaign was to eradicate the practice of open defecation by 2017.

By 2014, nearly half the population of India defecated in the open; that is, 90% of all those who defecated in the open in South Asia were from India. The primary cause for open defecation is the lack of access to toilets. However, lack of toilets is not the only impediment to sanitation. To make an area open defecation free (ODF), one must address the issue of the unavailability of sanitary toilets as well as the cultural sentiments surrounding toilet usage (Coffey et al. 2017). Thus, it is widely felt that to achieve ODF communities, one must address both issues: the supply-side constraints in terms of toilet availability and the demand-side bottlenecks regarding removing cultural attitudes towards using toilets. To accelerate the efforts for achieving universal sanitation coverage, the government of India launched the Swachh Bharat Mission (SBM) on October 2, 2014. Within the SBM framework, guidelines recommend the use of community-led approaches to motivate households to construct sanitary toilets and to deploy community-based monitoring and vigilance activities to create peer pressure on outliers to stop open defecation. To assist households living below the poverty line, the government provides a financial incentive of up to INR 12,000 per household to construct a sanitary toilet. Sanitation has huge network effects, implying that individuals will have higher chances of constructing and using toilets if they know that others in the community are doing so. This means the communication must be at a community level. Data, therefore, has an important role in ensuring sanitation outcomes.

In Chapter 19, Prerna Mukharya shows how data is used to understand why some states have fared better in SBM than others. The study analyzes the trends in sanitation and identifies the inflection points with respect to policy change, local factors, agencies, and other change-makers in the field of sanitation. The chapter then explores the primary reasons behind what, how, and why the major changes in the sanitation sector occurred in the five states and analyzes the impact on health indicators. The authors find that community-based action, government efficiency, panchayati raj, and village councils play a crucial role in driving CLTS.

In Chapter 21 in this section, Saloni Goel presents an interesting case where real-time data monitoring led to better sanitation outcomes. They study the role of data-driven decision-making in ensuring total sanitation for the Kumbh Mela in 2019. Kumbh Mela 2019 was organized in Prayagraj and witnessed a footfall of 240 million visitors over the 50-day duration. In this Kumbh, on the most auspicious bathing day of February 4, 2019, the number of visitors peaked at 30 million. The number of toilets to be provided had to be increased manifold to ensure an open-defecation–free festival. The government of Uttar Pradesh decided to ramp up the number of public toilets for Kumbh 2019 to almost 80,000. The authorities agreed to allocate space for public toilets along roadsides at frequent and regular intervals for easy visibility and access. However, some challenges remained, starting from toilet design to odor management to regular monitoring of the toilets. The main big data intervention came in the areas of maintenance and services, with an IoT-based smart toilet monitoring system designed and implemented for the main event. IoT was selected as the preferred tool because not only was it the only option available for real-time feedback, it was also cost effective over the scale and spread of the Kumbh. IoT also allows for complete transparency and reliability in two-way communication. Finally, it is also most cost effective. Further, a mobile device was given to the monitoring team, an essential prerequisite for IoT-based monitoring, which would raise the enthusiasm level. The immediate and most visible impact of this work has been of transforming Swachh Kumbh, an ambitious and incredible vision, into reality. Kumbh 2019 was declared open defecation and odor-free and culminated in the award of a Guinness World Record in sanitation. As the Mela progressed, reports of cleanliness published in national and international media became the biggest reason for drawing visitors, making Kumbh 2019 the most visited and inclusive event ever.

The final section combines studies on smart city and urban development.

0.7 SMART CITY AND URBAN DEVELOPMENT

Urban areas in India are vital centers of economic activity and cultural diversity, attracting millions in search of better opportunities and lifestyles. However, they face challenges like congestion, pollution, inadequate infrastructure, and social inequalities. The growth of urban areas necessitates robust urban planning, infrastructure development, and housing solutions to ensure sustainable and inclusive cities. Urbanization is closely linked to economic development, job creation, and increased living standards. Balancing the opportunities and challenges of urbanization is a significant task for the Indian government, emphasizing the need for efficient governance, infrastructure investment, and sustainable urban development practices to create thriving, resilient, and equitable urban spaces.

Smart cities are a crucial urban development initiative in India, reflecting a shift towards more sustainable and efficient urban areas. The government envisions these cities as technology-driven hubs that offer citizens improved infrastructure, services, and quality of life. By leveraging technology and data, smart cities aim to enhance urban mobility, reduce environmental impact, ensure efficient resource use, and promote citizen participation. These initiatives address the growing challenges of rapid urbanization, such as congestion, pollution, and resource constraints by fostering innovation and modernizing urban governance. Smart cities transform India's urban landscape and drive economic growth.

The SDGs related to smart city and urban areas covered in this section of the book are:

- Goal 8: Decent work and economic growth
- Goal 9: Industry, innovation, and infrastructure
- Goal 11: Sustainable cities and economies

Finally, this book's last section examines emerging and futuristic issues, focusing on smart cities and developing urban facilities. The section has a collection of chapters that use innovative techniques combining various forms of unstructured data. The first chapter in this section demonstrates how drone technology is used for early detection of road quality degradation, thereby fixing it to save costs. The next chapter demonstrates how an intelligent traffic management system (ITMS) was developed and implemented in Ahmedabad to control unruly traffic and minimize traffic violations at major junctions in the city. The third chapter shows how Google data is used to plan airports for cities by analyzing survey amplification techniques to determine affluence and thus the intent to travel to relevant cities.

Transport demand in most Indian cities has increased substantially due to an increase in population as a result of both a natural increase and migration from rural areas and smaller towns. The availability of motorized transport, increase in household income, and increase in commercial and industrial activities have further added to it. Reducing traffic congestion, vehicular emission, and accident risk requires a comprehensive strategy. The main objective of such a strategy should be to provide and promote sustainable, high-quality links for people, goods, and services to, from, and within the city. The strategy should be designed in such a way that it reduces the need to travel by personalized modes and boosts the public transport system. This requires not only increasingly stringent emissions standards, specifications for clean fuels, proper maintenance of in-use vehicles, optimal pricing of transport services, and demand- as well as supply-side management measures but also a complete overhaul of the public transport system. Poor road quality not only leads to traffic congestion, more commuting time, and increased pollution, but it also often leads to more road accidents. Though, the number of accidents is certainly on the higher side if one were to consider that the average vehicular speed in India is much lower than the global mean. According to a recent study by IMF, with a mean speed of 58 km/h, India is ranked 127th out of 162 countries as of 2022.[8] According to the National Crime Record Bureau (NCRB) data, in 2021, the 19 major metros (those with a population exceeding 1 million) witnessed a total of 6794 road accident–related deaths, which accounts for approximately 6 per lakh.[9] However, some of the more populous cities (Mumbai, Delhi, Kolkata) witnessed far lower numbers than the national average. This was partly because of the usage of technology and the effective usage of data techniques to tackle potential problems.

In Chapter 23 of this section, Jayavardhan Sambedu demonstrates how drone technology is used for early detection of road quality degradation and thereby fixing it to save costs. This chapter describes an innovative pilot approach in collaboration with the government of Andhra Pradesh where drones were flown over the road to capture images of the road and the depth of the potholes was measured using the onboarded LiDAR system. Machine learning methods were used to classify the road segments into good, average, bad, or mud roads, along with information on the line markings. This information, along with the dimensions of the potholes, was reported in digital format to state authorities for real-time monitoring of the road condition and taking up repair activities. In Chapter 25, Juzer Bharmal and Gayatri Srinivasan demonstrate how an intelligent traffic management system was developed and implemented in the city of Ahmedabad to control unruly traffic and minimize traffic violations at major junctions in the city. AI-based cameras were fitted upon over 60 traffic junctions to detect traffic violations. The camera would snap evidence images at high speed, automatically pick up the number plate, access the vehicular database, and identify vehicle owners. The challan issuing process was automated as well. The Intelligent Traffic Management System (ITMS) generates over 1 Lakh e-challans a day for over 60 junctions, with the police dispatching at least 30,000 challans (15 times more challans) every day. This has led to better discipline on the roads of Ahmedabad. Other benefits include tracking lost or stolen vehicular data and tracking vehicles that are involved in criminal activities. In this way, artificial intelligence supports the police in maintaining order in the city.

Another key issue that often plagues rapid urbanization in India is keeping cities safe in terms of the prevention of crime.[10] With only 67% chargesheet rates, the problem is further compounded as fewer crimes are reported, given such low chargesheet rates. However, data analytics can substantially ameliorate the problem by helping in crime detection and prevention. In Chapter 24, Ayushi demonstrates how AI ML enabled a state-of-the-art real-time surveillance system for monitoring and security in cities like New Delhi, Chennai, and Surat. The intervention involves cameras equipped with the latest facial recognition technology that capture images in real time and relay them back to a huge screen installed at the command center. Speed and facial matching are the biggest features of the system. Since the installation of the system, Surat has reported a 27% drop in crimes, which is a huge achievement for the police and city administration.

The last couple of decades have seen fantastic growth in air traffic, and with growing urbanization, newer airports are needed. In Chapter 26, Akhilesh Tilotia shows how alternative big data (e.g. Google search queries) is used to plan airports for cities. Millions of anonymized and aggregated data points were collected in the form of travel-related search queries, the complexity and diversity of business in cities, and general search queries like travel and news for various cities in India. These data points were then analyzed using survey amplification techniques to determine affluence and intent to travel for relevant cities. By combining the diversity-complexity-Human Development Index parameters, eight cities were found to have an independent potential of attracting 0.5 million passengers. Twenty of the 30 cities suggested were picked by the government for the next Ude Desh ka Aam Naagrik (UDAN) bidding scheme.

We hope that readers will enjoy these case studies as much as we did while compiling them. The studies are contemporary and will certainly pave the road towards Viksit Bharat using data science.

NOTES

1 https://sdgs.un.org/goals
2 https://thewire.in/rights/world-population-day-withering-demographic-dividend#:~:text=India%20entered%20the%20demographic%20dividend,for%20the%20demographic%20dividend%20window
3 https://www.orfonline.org/expert-speak/encashing-indias-demographic-dividend/
4 https://www.iea.org/reports/india-energy-outlook-2021/energy-in-india-today
5 https://www.ibef.org/industry/power-sector-india
6 https://www.adriindia.org/adri/india_water_facts
7 https://pib.gov.in/PressReleaseIframePage.aspx?PRID=1604871
8 https://www.hindustantimes.com/india-news/how-fast-is-road-transport-in-india-101655442867532.html
9 https://morth.nic.in/road-accident-in-india
10 https://ncrb.gov.in/en/Crime-in-India-2021

BIBLIOGRAPHY

Ahluwalia, M. S. (2020). *Backstage: The Story Behind India's High Growth Years*. Rupa Publications.

Balarajan, Y., Selvaraj, S., & Subramanian, S. V. (2011). Health care and equity in India. *The Lancet*, 377(9764), 505–515. https://doi.org/10.1016/S0140-6736(10)61894-6

Banerjee, A., & Duflo, E. (2011). *Poor Economics: A Radical Rethinking of the Way to Fight Global Poverty*. PublicAffairs.

Coffey, D., Gupta, A., Hathi, P., Khurana, N., Spears, D., Srivastav, N., & Vyas, S. (2014). Revealed preference for open defecation: Evidence from a new survey in rural North India. *Economic and Political Weekly*, 49(38), 43–55.

Coffey, D., Spears, D., & Vyas, S. (2017). Switching to sanitation: Understanding latrine adoption in a representative panel of rural Indian households. *Social Science & Medicine*, 188, 41–50.

Dickin, S. K., Bisung, E., & Nansi, J. (2021). Empowerment in water, sanitation and hygiene index. *World Development*, 137, 105158. https://doi.org/10.1016/j.worlddev.2020.105158

Donoho, D. (2017). 50 years of data science. *Journal of Computational and Graphical Statistics*, 26(4), 745–766. https://doi.org/10.1080/10618600.2017.1384734

Drèze, J., & Sen, A. (2013). *An Uncertain Glory: India and Its Contradictions*. Princeton University Press.

Dye, T. R. (2017). *Understanding Public Policy* (15th ed.). Pearson.

Griggs, D., Stafford-Smith, M., Gaffney, O., Rockström, J., Öhman, M. C., Shyamsundar, P., . . . & Noble, I. (2013). Policy: Sustainable development goals for people and planet. *Nature*, 495(7441), 305–307. https://doi.org/10.1038/495305a

Hanushek, E. A., & Woessmann, L. (2008). The role of cognitive skills in economic development. *Journal of Economic Literature*, 46(3), 607–668. https://doi.org/10.1257/jel.46.3.607

Howlett, M., Ramesh, M., & Perl, A. (2009). *Studying Public Policy: Policy Cycles and Policy Subsystems* (3rd ed.). Oxford University Press.

IEA. (2021). *India Energy Outlook 2021*. International Energy Agency, Paris. https://www.iea.org/reports/india-energy-outlook-2021

IPCC. (2022). *Climate Change 2022: Impacts, Adaptation and Vulnerability*. Contribution of Working Group II to the Sixth Assessment Report. Cambridge University Press. https://doi.org/10.1017/9781009325844

Janssen, M., Charalabidis, Y., & Zuiderwijk, A. (2012). Benefits, adoption barriers and myths of open data and open government. *Information Systems Management*, 29(4), 258–268. https://doi.org/10.1080/10580530.2012.716740

Jordan, M. I., & Mitchell, T. M. (2015). Machine learning: Trends, perspectives, and prospects. *Science*, 349(6245), 255–260. https://doi.org/10.1126/science.aaa8415

Kamilaris, A., Kartakoullis, A., & Prenafeta-Boldú, F. X. (2017). A review on the practice of big data analysis in agriculture. *Computers and Electronics in Agriculture*, 143, 23–37. https://doi.org/10.1016/j.compag.2017.09.037

Kapur, D., Mehta, P. B., & Vaishnav, M. (Eds.). (2017). *Rethinking Public Institutions in India*. Oxford University Press.

Kelkar, V., & Shah, A. (2022). *In Service of the Republic: The Art and Science of Economic Policy* (2nd ed.). Penguin Allen Lane. (First edition 2019)

Khera, R. (Ed.). (2011). *The Battle for Employment Guarantee*. Oxford University Press.

Kitchin, R. (2014). The real-time city? Big data and smart urbanism. *GeoJournal*, 79(1), 1–14. https://doi.org/10.1007/s10708-013-9516-8

Kohli, A. (2012). *Poverty Amid Plenty in the New India*. Cambridge University Press.
Kummu, M., Guillaume, J. H. A., de Moel, H., Eisner, S., Flörke, M., Porkka, M., . . . & Ward, P. J. (2016). The world's road to water scarcity: Shortage and stress in the 20th century and pathways towards sustainability. *Scientific Reports*, 6, 38495. https://doi.org/10.1038/srep38495
Lafortune, G., Fuller, G., Moreno, J., Schmidt-Traub, G., & Kroll, C. (2018). *SDG Index and Dashboards Report 2018: Global Responsibilities*. Bertelsmann Stiftung and SDSN, New York. https://www.sdgindex.org/
Lobell, D. B., Schlenker, W., & Costa-Roberts, J. (2011). Climate trends and global crop production since 1980. *Science*, 333(6042), 616–620. https://doi.org/10.1126/science.1204531
Marda, V., & Narayan, S. (2020). Data governance in the public sector: Lessons from India. *Big Data & Society*, 7(2), 1–14. https://doi.org/10.1177/2053951720970555
Mohanty, S. P., Choppali, U., & Kougianos, E. (2016). Everything you wanted to know about smart cities: The internet of things is the backbone. *IEEE Consumer Electronics Magazine*, 5(3), 60–70. https://doi.org/10.1109/MCE.2016.2556879
Nilekani, N. (2008). *Imagining India: The Idea of a Renewed Nation*. Penguin Press.
NITI Aayog. (2017). *Progress On Drinking Water, Sanitation and Hygiene: 2017 Update And SDG Baselines (PDF)*. NITI Aayog.
Pellissery, S., Sidhwani, P., & Chatterjee, B. (Eds.). (2021). *Public Policy and Politics in India: How Institutions Matter*. Oxford University Press.
Provost, F., & Fawcett, T. (2013). *Data Science for Business: What You Need to Know About Data Mining and Data-Analytic Thinking*. O'Reilly Media.
Psacharopoulos, G., & Patrinos, H. A. (2018). Returns to investment in education: A decennial review of the global literature. *Education Economics*, 26(5), 445–458. https://doi.org/10.1080/09645292.2018.1484426
Sabatier, P. A. (Ed.). (2007). *Theories of the Policy Process* (2nd ed.). Westview Press.
Sachs, J. D. (2012). From millennium development goals to sustainable development goals. *The Lancet*, 379(9832), 2206–2211. https://doi.org/10.1016/S0140-6736(12)60685-0
Sachs, J. D., Lafortune, G., Fuller, G., & Drumm, E. (2023). *Implementing the SDG Stimulus*. Sustainable Development Report 2023. Dublin University Press/SDSN. https://dashboards.sdgindex.org/
Sinha, A. (2005). *The Regional Roots of Developmental Politics in India: A Divided Leviathan*. Indiana University Press.

Section I

Planning and Governance

1 CM Relief Fund prediction – The Right Amount of Funds at the Right Time for the Right Cause

Anand Kulkarni and Raunak Saraf

ABSTRACT

The Chief Minister's Relief Fund (CMRF) provides relief to the state's citizens in distress due to calamities or diseases by providing financial assistance to people affected by floods, droughts, and accidental or natural disasters. A shortage of funds means several requests are not fulfilled, and excess funds would require further justification from the donors. The CM Dashboard has provided a solution to all these issues in the form of a forecast data analysis model. The forecast data analysis model calculates the existing and required funds and understands the utilization and provision of funds for social welfare in a simple, clean, and navigable manner. In this way, the CM Dashboard helps the officers responsible get a good picture of the current situation of funds disbursed across various factors such as time (month), diseases, and districts.

> Profitability should not only mean surplus over costs. It should also mean the provision of investment for future growth.
>
> **– Chanakya**

1.1 INTRODUCTION

Like in any other country in the world, in India, the loss of life or property owing to various shocks, natural calamities, or diseases is immense. Good governance is crucial in minimizing this damage and compensating for the losses incurred. And, of course, how do we compensate for the loss caused?

The Chief Minister's Relief Fund (CMRF), a trust registered under the Bombay Public Trusts Act, 1950, is managed by the state of Maharashtra under the chairmanship of the honorable chief minister, who exercises overall supervision and control of this fund. The funds in the CMRF are utilized to relieve the needy elements of society who are the victims of natural calamities or critical diseases. CMRF aims to provide immediate relief to such state citizens who are in distress due to calamities or diseases. CMRF also offers financial assistance to people affected by floods, droughts, and accidental or natural disaster–related fires.

Briefly, funds under CMRF are utilized for the following purposes:

- **Natural calamities**: Help for natural calamities in the state of Maharashtra and other states across the country.
- **Victims of communal riots or terrorist attacks**: Financial help to the heirs of persons who have died in communal riots or suffered injury or loss to their property.

DOI: 10.1201/9781003487531-3

- **Surgery for critical diseases**: Final help for surgery and/or treatment for patients suffering from critical diseases.
- **People who die in accidents**: Financial assistance or any other help to the heirs of people who die in accidents (excluding motor, railways, airplane, or ship accidents).
- **Needy institutions**: Financial assistance to various institutions in need of financial or other help.
- **Educational, social, and cultural seminars or gatherings**: Financial assistance or any other help for organizing educational, social, and cultural seminars or gatherings.
- **Construction of educational or medical buildings**: Financial assistance or any other assistance for constructing buildings for educational or medical establishments.

To address some of these shocks, this fund has a budgetary contribution of INR 1 Crore by the CM of Maharashtra. While a large part of the fund is used to address citizen distress, some charitable organizations also get periodic financial assistance from CMRF. All financial assistance is processed and disbursed at the Department of General Administration.

1.2 MANAGING CMRF AND ITS CHALLENGES

Manual observation and calculations of existing and required funds were some of the most practiced traditional methods for the funds' distribution for CMRF. For example, state officials used to visit affected locations, collect physical data on the ground, and report the status. However, in some cases, for example, in the case of critical illnesses such as cancer, the officers would get data from the authorized sources, calculate the required amount, and then provide funds. Inefficiencies in the system gradually appeared.

It was turning out to be a massive challenge to keep track of donations and estimate a budget for the funds in CMRF because of the unstable nature of the financial requirements. The unpredictability of the budgetary allocation, disbursements, and their subsequent utilization was proving a major obstacle for the Maharashtra government and hindered the smooth operation of the fund.

At the outset, it must be noted that the state government is responsible for properly operating the CMRF. Having too little in the fund would lead to less disbursement for the needy, but having excess funds would also pose a challenge. If the funds remaining in CMRF are more than expected, the government is answerable to the donors about why those funds are not being allocated. Of course, the perennial question that the government must answer to its own satisfaction and that of the citizens is whether the funds being disbursed are properly utilized.

For these reasons, the government was in search of smart solution that could offer a bird's-eye view of the existing funds, overall requirements of society, and required funds' disbursement. Though the contribution from the CM office is predefined, the allocation of funds under various heads often is stochastic and requires accurate prediction well in advance. In this circumstance, planning and disbursement of the available funds became crucial for needy ones at critical times of their illness and hospitalization, and so on.

Thus, there was an intense requirement for a strong information reporting framework to make smart decisions for the CMRF.

1.3 CRITIQUE OF THE EXISTING APPROACHES

One of the major challenges was in cases where, after estimating the budget and distributing funds, there was a requirement for more funds. The main reason behind this was that a post-factor analysis was required after the funds' disbursement, which wasn't in place. For example, funds were provided to a needy person who underwent surgery for cancer. However, due to an unexpected criticality, the type of surgery or the required treatment had to change. In such cases, the allotted fund turned out to be insufficient. Thus, a smart solution was what the government decided upon.

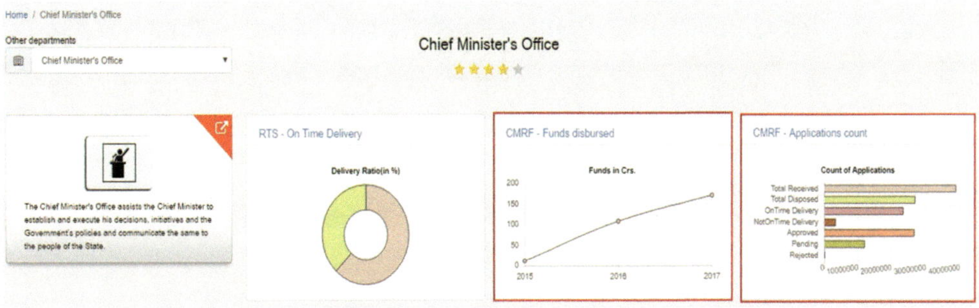

FIGURE 1.1 CM Relief Fund Dashboard overview.

The Maharashtra IT Corporation Limited contacted various departments and collected details about the CMRF. Keeping these in mind, they created smooth algorithms that understand and execute the programs in a dynamic environment. Further, all data received was cleaned and then transformed to be accurately displayed in the dashboard.

1.4 THE NEW APPROACH FOR MANAGING CMRF

The CM Dashboard has solved all the mentioned issues through a *forecast data analysis* model. The data analysis solution has provided a scientific method to estimate the budget of the CMRF. Specifically, the CM Dashboard provides an analytical platform for data analysis (Figure 1.1). State officials can now analyze and use accurate real-time data and make further decisions accordingly.

The forecast data analysis model used in CMRF utilization is at the core of this success story. The forecast model calculates the existing and required funds and understands the utilization and provision of funds for social welfare in a simple, clean, and navigable manner. In this way, the CM Dashboard helps officers understand the current situation of funds disbursed across various factors such as time (month), diseases, and districts.

The high-level dashboard, which the chief minister monitors regularly, shows a snapshot of the total funds disbursed and details of the requests the CM office has received for funds. The dashboard further provides a breakdown of application requests in terms of applications processed, applications approved, and cases where the funds have been disbursed (Figure 1.2).

1.5 HOW DOES THE CM DASHBOARD WORK?

When a department user opens a scheme-specific page in the Analytics Dashboard through the CM Dashboard, they can view the forecasting of the required funds. Based on the forecasts offered by the dashboard, the officials can figure out the approximate amount to be distributed to individuals based on their needs. This information is granular (individual level as well as at the district level) and is updated monthly. The forecast is based on the past historical data for several years of the fund requests received and those that were fulfilled, disbursed, and rejected by the CM office – this historic data forms a good basis for time series forecasting of the future period. A snapshot of the forecast for the year 2019 is given in Figure 1.3.

Once the dashboard is in place, the key performance indicators (KPIs) are categorized into situational and department reports. KPIs are defined for the department and each of its initiatives. Daily/monthly status reports and graphical reports are also offered on the CM Dashboard. Indicators such as filtering and drill-down are available in the dashboard and contain the lead and lag performances district-wise, and so on. For clarity, districts leading in the KPIs given to them are ranked as the lead performance districts, whereas those unable to fulfill the designation KPIs are shown as lag performance districts.

FIGURE 1.2 CM Relief Fund Dashboard details.

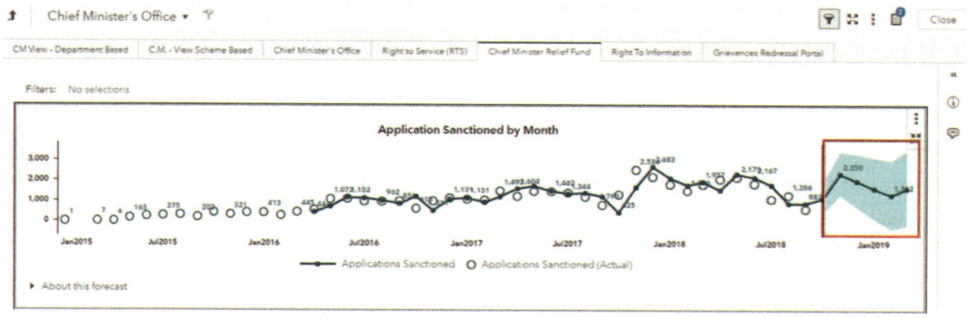

FIGURE 1.3 Fund Forecast on CM Relief Fund Dashboard.

Another important aspect is that the data is automatically updated without any assistance from the user. We wanted to give discretionary powers to the department officials, and they can opt to set the frequency of the data updating. Therefore, the dashboard's frequency of data updating will vary with departments per their indicated choices. The additional facility of SMS alerts to higher authorities on timely data update frequency is also provided. The dashboard further has search options, that is, ways to filter reports and customize search results to find exactly what a user want, and a ranking of districts based on their performance in terms of predefined parameters has been implemented to create healthy competition among districts.

1.6 DATA TYPE USED

The data collected was in the form of textual, numerical, and metadata. The district-wise data for multiple incidents for previous years was also provided by concerned government departments. All this data is used to create a forecast model. This predictive data analysis model forecasts funds and requests in the coming months for each district for a predefined set of diseases based on system-generated parameters.

Specifically, data is retrieved from the various data centers (DC) such as the State Data Center (SDC) and National Informatics Center (NIC) by establishing virtual private network (VPN)

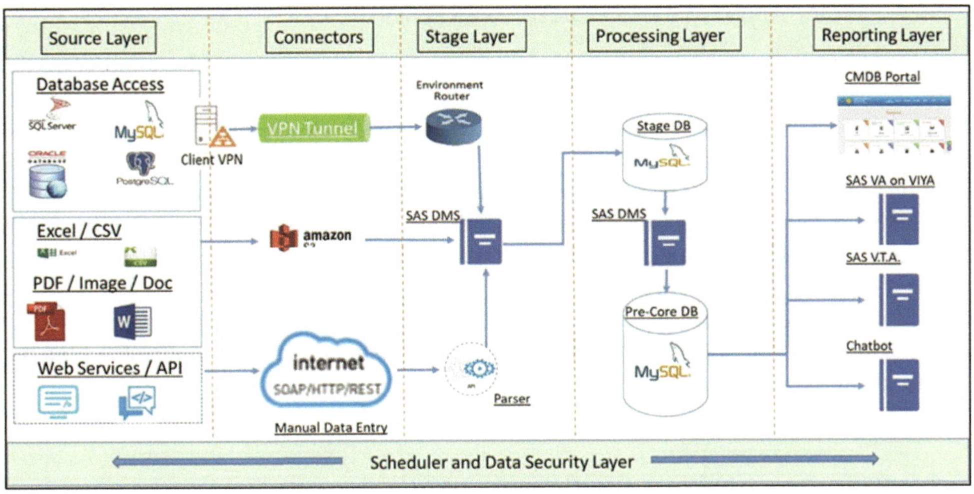

FIGURE 1.4 Data processing architecture of the CMRF application.

connectivity between the hosted cloud site and the data center sites. After connecting these sites, the Data Integration module retrieves data by creating an Open Database Connectivity (ODBC) connection with the database available in the DC. For this purpose, a separate ODBC connection is created for each database of the DC.

Typically, the solution for processing data is provided in the following layers (Figure 1.4): source, connectors, stage, processing, and reporting.

1.7 DATA COLLECTION AND PROCESSING – CHALLENGES AND SOLUTIONS

The departments and CM office were looking for a unified application to collect and analyze data from various sources, analyze data using the latest data science methods, and easily disseminate actionable inputs for policymakers and government officers.

As part of the data collection process, various departments identify their respective KPIs and decide the frequency of the data sharing (daily/weekly/monthly/yearly). After that, the department decides the mode of primary data sharing (manual/dynamic/interface/redirection) and the mode of sharing for the next level of data (redirection to the parent department system or display in the CORE itself, etc.).

The department provides the relevant web services and data ingesting system in the case of dynamic updating. It develops the KPI interface per the template and integrates it with analytics to display the data in interface updating mode.

Such an approach may be extended across different departments to derive actionable insights specific to such departments. The various steps in the process are highlighted in the process diagram. Governments collect a large amount of administrative and operational data, which usually lies in silos across flat files or disconnected databases – collecting this data and segregating it into a single database is the first step in this process.

As these datasets are from disparate sources, quality check and pre-processing of the data form the next important step of this process. The next step consists of generating dashboards on top of the data, which requires defining the key performance indicators in discussion with the departments. Analysis of these dashboards along these KPIs helps in the identification of the root causes and areas for improvement – this information can thus be passed in real time to the officers in charge of these activities through SMS or email.

1.8 SOLUTION AND TYPE OF DATA ANALYSIS

Maharashtra IT Corp Ltd and experienced statisticians have come up with the solution for the data analysis of CMRF. The simple exponential smoothing (SES) method for time series forecasting is used (Figure 1.5). Simple exponential smoothing is a basic forecasting technique for univariate time series data with no trend or seasonality. It predicts future values based on a weighted average of past observations. The weights decrease exponentially as observations get older, giving more importance to recent data. This model forecasts the funds for various applications and utilization areas. These amounts would further help stakeholders plan for the required funds for listed departments in various districts of Maharashtra.

The following is the gist of CMRF-analyzed data in the CM Dashboard per application sanction by month, number of donors, and amount of donation. This analysis helps understand the requirement of funds in the year 2019 in various districts.

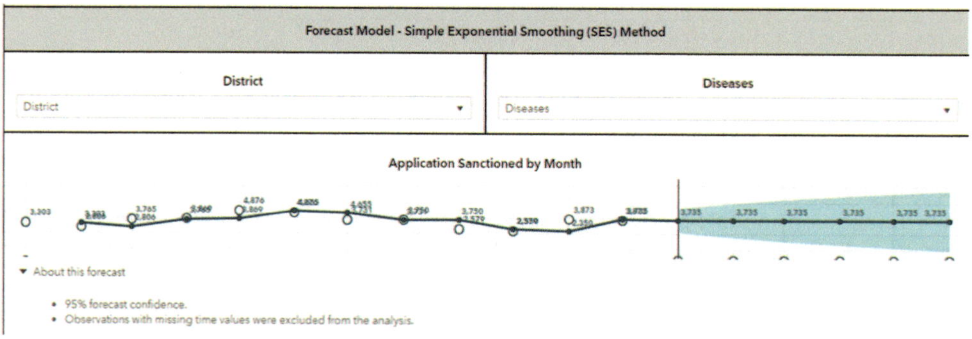

FIGURE 1.5 Simple exponential smoothing for forecasting future values.

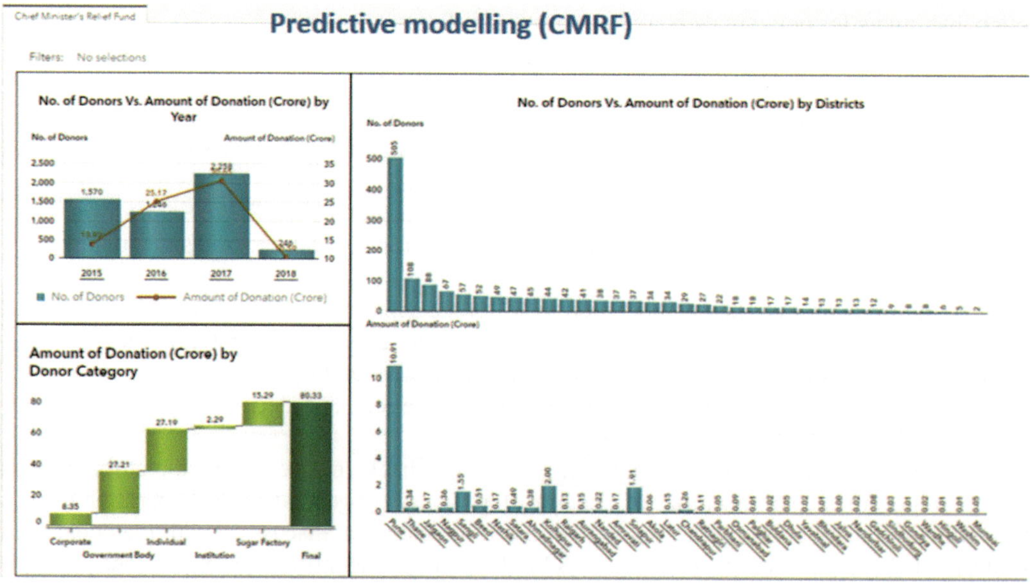

FIGURE 1.6 Details of donation across time and districts.

1.9 OUTPUT

The funds requested are forecast in the model for various incidents or diseases for the next couple of months, giving sound data-based advice to stakeholders when they must decide on the disbursement of funds. The analysis dashboard also helps government officials get a single view of the funds in terms of the donations received in CMRF over the period, utilization of the funds across sectors, and utilization of the funds across different districts of Maharashtra (Figure 1.6). This provides adequate information for making quick decisions regarding the disbursement of funds in times of need and reducing procedural delays.

Finally, the aim of the government is that the right amount of funds be utilized at the right time for the right cause using this predictive data analysis model, which is achieved through the CMRF Dashboard. In some cases, complete accuracy for fund distribution is achieved, which gives immense satisfaction to the Maharashtra government.

Currently, the CM Dashboard portal is helping in better data forecasting, leading to better budgeting of government funds and optimizing their utilization per month, per district, and per requirement. The cascade effect will be that people needing funds are helped as efficiently and effectively as possible.

BIBLIOGRAPHY

Bennett, Roger, and Rita Kottasz. "Emergency fund-raising for disaster relief." *Disaster Prevention and Management: An International Journal* 9, no. 5 (2000): 352–360.

Mukherjee, Sacchidananda. *State of public finance and fiscal management in India during 2001–16*. National Institute of Public Finance and Policy, 2019.

Sharma, Vinod K. "Natural disaster management in India." *Indian Journal of Public Administration* 43, no. 3 (1997): 764–774.

Wildasin, David E. "State and local government finance in the current crisis: Time for emergency federal relief?" *Municipal Finance Journal* 29, no. 4 (2009): 49–64.

2 Sustainable Development Goals: India Index
The Journey of Creating It

Anjali Bansal

ABSTRACT

The SDG India Index tracks the states' progress on sixteen SDG parameters like health, education, economic growth, climate change, and environment. This chapter delves in the process of transforming the national-level SDG indicators into state-level indicators that would appropriately measure the SDG, followed by the challenges involved in acquiring recent data to create the index, and finally discusses the process involved in creating the first state SDG index for India.

2.1 INTRODUCTION

As of 2017, out of 195 countries recognized by the United Nations, there are only 31 developed countries. This means that only 15% of countries are presently developed globally. Alternatively, it means that 85% of countries are either developing or emerging. It signifies that collaborative and collective development actions need to be taken for rest of the world. Embracing the need to have a shared long-term development action agenda for the prosperity of the whole of humankind, the United Nations developed the Millennium Development Goals (MDGs) in 1990 to galvanize all countries and communities into action. The MDGs consist of eight international development goals with 18 quantifiable targets. In the MDG era, the world experienced significant economic changes. India was also one of them. Unfortunately, the implementation of the MDGs in India was not up to the mark. India has made distinguished progress in achieving some MDGs, like Goal 1 of reducing poverty and Goal 3 of increasing enrolment at primary and secondary schools, but because of a lack of proper monitoring, evaluation, and accountability, the achievement of MDGs has not been properly widespread. Later, it was realized that these limitations could be overcome if the government had a proper data monitoring system. It is not wrong to say that one of the major reasons for the unsatisfactory performance of MDGs was the lack of data and obsolete data collection mechanisms. I here want to quote Daniel Keys Moran, a famous writer and computer programmer who emphasized the importance of having data by saying, "You can have data without information, but you cannot have information without data." Eventually, it was realized that proper data and its usage is essential.

In 2015, when Sustainable Development Goals (SDGs) came into existence, this time, the country's approach was different. Taking lessons from the failure of MDGs, India is now better prepared and equipped to participate and showed its capacity to achieve the 2030 goals. The SDGs are a bold universal set of 17 goals and 169 targets ranging from poverty eradication, human health, and sanitation to urban settlements and safeguarding the global ecosystems on which humanity depends for its survival. They were formed to organize and streamline development actions for the greater achievement of human well-being. Under the leadership of our hon'ble prime minister, India is determined to implement the development agenda of "leaving no one behind" successfully. The day the SDGs came into existence, a mandate was given to the Ministry of Statistics and Programme Implementation (MoSPI),

DOI: 10.1201/9781003487531-4

the National Statistical Unit, and NITI Aayog, the Government of India policy think tank, to oversee the implementation of SDGs in the country. This time, the aim was to reap the advantage of the SDG era.

In a federal system like India, where districts build states and states build the nation, it is one of the challenging tasks for the central government to oversee the progress of the whole country across all parameters simultaneously. Moreover, India is the home of one-sixth of the world's population, living across 29 states and 7 union territories (UTs). Every state has its unique opportunities and challenges. Some are blessed with enormous natural resources, and some are endowed with world-class infrastructure and business ecosystems. There is no matrix system available in our country that keeps a record of these things. In our country, we cannot have one policy/scheme and programs that fit across all states. Therefore, it is a country with some of the most ambitious developmental and social schemes and programs. It is clearly recognized that the development of our country cannot be ensured by launching a single scheme. Therefore, the SDGs are like one colossal umbrella program that covers all relevant dimensions. Through the SDGs, a common matrix has been built, which at least gives insight into the current position of states/UTs.

2.2 THE INCEPTION OF THE IDEA OF BUILDING A COMPOSITE SDG INDIA INDEX

NITI Aayog is mandated to coordinate SDG work by adopting a synergistic approach involving union ministries and departments, state governments, administration of UTs, civil society organizations, academia, and the business sector to achieve India's SDG targets. The instructions were clear that this time: awareness of SDGs should reach the local level, unlike the MDGs, which remained only in the reports of the government. Initially, a lot of consultations and meetings were organized with all the relevant stakeholders mentioned previously. Still, we cannot finalize the actions needed to accomplish the task. Finally, taking lessons from some developed countries, it was decided that India would build its baseline SDG India Index Report–2018. Given the importance accorded by the government of India to achieving the SDGs, NITI Aayog decided to estimate the progress through a single measurable index that would serve as an advocacy tool and trigger action at the state level.

2.3 ADVANTAGES OF THE SDG INDEX

NITI Aayog has the twin mandate to oversee the implementation of SDGs in the country and promote competitive and cooperative federalism among states and UTs. The SDG India Index bridges these mandates, aligning the SDGs with the five Ps of the global SDG movement – people, planet, prosperity, partnership, and peace. The SDG India Index is intended to provide a holistic view of the social, economic, and environmental status of the country and its states and UTs. The SDG Index can be useful to states/UTs in assessing their starting point on the SDGs in the following ways:

i. Supports states/UTs to benchmark their progress against the national targets.
ii. Boosts states to compete with their peers and devise better strategies to achieve the SDGs by 2030.
iii. Supports them to identify priority areas, as working on all SDG areas at the same time may be challenging.
iv. Helps them to modulate their initiatives and investments as well as to measure incremental progress.
v. Highlights data gaps across SDGs and identifies areas for building individual and systems capacity for better data management.

2.4 PROCESS OF CHOOSING INDICATORS AND CHALLENGES FACED

Each of the 17 Sustainable Development Goals is mapped with a set of 169 targets to be achieved by 2030. The United Nations has framed a Global Indicator Framework. Every country can tailor-make its National Indicator Framework. This framework was used as the foundation for building the SDG India Index. In India, the MoSPI has constructed a National Indicator Framework comprising 306 indicators. NITI Aayog has comprehensively mapped SDG targets with GoI schemes and programs, focusing on their interlinkages. It regularly consults multiple stakeholders on different goals, capacity building, evaluation framework, knowledge dissemination, and progress mapping.

Three hundred six indicators are too many to build an index. Moreover, data was also not available for all 306 indicators. At times, something is better than nothing. Going by this saying, NITI Aayog decided to prioritize some national indicators out of the national indicator framework, on which state-wise data is available. The central government intends to improve the well-being of states by boosting competition, and this led to the unique idea of constructing the SDG India Index. The process of selecting the priority indicators was very consultative. At first, NITI Aayog chose around 75 indicators out of 306. Then, after consultation meetings with 38 central ministries/departments, MoSPI and state governments were organized. After long discussions in meetings with concerned central ministries regarding priority indicators, 62 indicators were selected. A rift was often created between the NITI Aayog and the related stakeholders in selecting priority indicators. The biggest challenge was to tap the state-level indicators and data. There are two types of central government schemes. One is the central sector scheme in which the center did 100% of the budget allocation. Another is centrally sponsored schemes in which the budget is shared in the agreed-upon ratio between the center and the state. Difficulty arises when the schemes are centrally sponsored. The center often might not have complete information on the overall performance of schemes in the states. Therefore, another important decision was made to avoid the difficulty of tapping data from states. It was decided that only those indicators whose data sources primarily emerge from central official statistical systems, that is, the respective union ministries/departments, would be selected.

Another limitation while deciding the indicator was a differentiation in states' characteristics. One type of scheme does not fit all. For example, a scheme like Mahatma Gandhi National Rural Employment Guarantee Act (MGNREGA) ensures 100 days of guaranteed employment in rural areas. States like Punjab do not exhaust the funds allocated under this scheme because they don't require that much fund allocation. But while constructing the index, because of this indicator, the position of Punjab came down. To avoid such problems, a reasonable choice of indicators should be made, and correct identification of common denominators is required. In this case, we finally decided to take the ratio of MGNREGA beneficiaries and number of the population living below the poverty line. Likewise, many such instances occurred where a lot of thinking was invested in getting the indicator that fit in all states.

It is well known that the biggest challenge in choosing the most suitable consensus indicator was the lack of data. Finally, NITI Aayog made a call that the SDG Index must be created despite any challenges. Along with this, it gave a cautious note that though the SDG India Index 2018 may not be a comprehensive representation of overall SDG dimensions since it is constrained by limited data availability, it would give an initial indication of states' positions. This index will help highlight each state's opportunities and challenges in all dimensions and supplement evidence-based policy-making by involving states/UTs. Despite these gaps and limitations, the SDG India Index 2018 provides critical insights into the status of SDGs in the country, although constrained by limited data availability. As data availability improves and new estimation techniques become available, subsequent reports of the SDG India Index will become more comprehensive with additional indicators and help measure incremental progress. After the huge step of selecting the priority indicators, the next major step is to finalize the SDG Index methodology.

2.5 PROCESS OF CHOOSING METHODOLOGY AND JUSTIFICATIONS

Many methodology options are available in the pool of tools and techniques. Here, we need to select the most appropriate methodology that converges well with the practical mechanism of all relevant stakeholders. While selecting the method, the most decisive step was choosing between a fixed and equal weighing strategy and flexible and unequal weighing strategy. Looking at the set-up of SDGs, the working mechanism of union ministries/departments, and the state governments, and after expert consultations, it was decided that all indicators under SDGs would get equal weight.

The justification was that in the first SDG report, every dimension, every indicator, and the ministries responsible for any indicators would be treated equally. Following an equal weightage strategy for all indicators across all goals has two limitations. First, it leads to biased results, and second, not all indicators are equally relevant to all states, which pulls down the performance of the state on those indicators. For example, schemes like the Mahatma Gandhi National Rural Employment Guarantee Scheme under SDG 1 of "No poverty" and Pradhan Mantri Gram Sadak Yojana under SDG 9 of "Better infrastructure" are less relevant in states like Punjab and Haryana because they have already been performing better on these issues. Thus, their progress on these schematic indicators will seem like less than that of other states, but it does not signify that these states are poor performers. Hence, it was decided that no extra burden or focus would be placed on any specific indicator. Ministries and equal weightage were assigned to all the priority indicators under the considered SDGs.

2.6 DATA COLLECTION PROCESS AND CHALLENGES

> Data is a precious thing and will last longer than the systems themselves.
>
> **– Tim Berners-Lee, father of the World Wide Web (November, 1990)**

Priority indicators were selected, and the methodology was decided. We have arrived at the most interesting and challenging part: data collection. The nature and source of data across all 62 indicators were very distinct.

Data on some indicators was very easily available, for example, MGNREGA scheme data from the real-time MGNREGA portal maintained by the Department of Rural Development, data related to health statistics from three rounds of the National Family Health Statistics survey conducted by the Ministry of Health and Family Welfare, and data on school education from the Unified District Information System for Education (U-Dise) developed by the National University for Educational Planning and Administration.

For some indicators, data was available in general but this availability was limited to the larger states only and at times missing for the northeastern states. For example, data on maternal mortality rates is available for only 22 major states.

There are some indicators for which data need to be collected directly from the concerned union ministries/departments. For example, for "number of auxiliary nurses and midwives (ANMs)", each state's midwives and doctor personnel must be collected from the Ministry of Health and Family Welfare. The proportion of the population with access to food grain at subsidized prices must be collected directly from the Department of Food and Public Distribution.

There are some indicators on which no data was collected at all. For example, data on solid waste management in urban areas was not collected. There are some indicators on which states have their respective data, but no aggregated data at the national level was available, for example, the number of seats women hold in national parliament, state legislation, and local self-government.

Apart from the data availability problem, there is another problem of data comparability. To avoid this problem in many indicators, a lot of time was invested in searching for the appropriate denominator. For example, the number of ATMs in states will give no meaningful information until or unless it is divided by an appropriate denominator. The appropriate denominator to gauge the state-wise coverage of ATMs is to show the number of ATMs per one lakh population.

The biggest challenge in data collection and availability has been witnessed in the environment-related dimension. The data problem was so peculiar that we had to drop SDG 12, 13 and 14 from SDG India Index calculation. In the journey of collecting data for the SDG index, we realized that our government needs to spend resources and involve expertise to revamp the data monitoring unit. Target-based decision making is needed, and the availability of good data is the foremost requirement. It highlighted the need to increase the capacity and capability of data collection in our country.

2.7 SDG INDIA INDEX 2.0

The SDG India Index in its second edition shows substantial refinement over its first edition. In the second edition of the report, the process of indicator selection and methodology assortment was more inclusive, peer reviewed, and holistic. This time, under the aegis of the Ministry of Statistics and Programme Implementation, a committee was formed with representatives of NITI Aayog, MOSPI, UN, and external experts to review the methodology. SDG Index 2.0 is more robust than SDG Index 1.0. Rankings of all goals will be done except SDG 17. In comparison to 62 indicators in the previous edition, this will have 100 indicators. Including 38 more indicators and the availability of state-wise data on them is a massive achievement. With this pace of achievement, the day is not far when we will have the SDG India Index with the 306 indicators that were initially defined in the national indicator framework. The SDG report is working as an effective tool to strengthen competitive federalism. States compare their progress on indicators and compete among themselves.

2.8 LOCALIZATION OF SDGS

Since the inception of SDGs in 2015, the first report on localization was launched in 2017-18, with updates to the same planned on an ongoing basis. Not only is there a state-level report; we have been successful in taking SDGs to the district level. In last two years, several state-level consultations were done to apprise states about the SDGs. NITI Aayog also prepared a northeastern SDG index report in which all the indicators were at district level. Out of 120 districts, the report included 103 districts across 84 indicators while constructing the index. As data plays a critical role in strengthening SDG localization, it is imperative to take adequate measures to build robust statistical systems at the center, state, and district levels. The day is not far off when India will have each state's SDG index.

2.9 CONCLUSION

The biggest motivation to kickstart the construction of the SDG India Index was that it is intended to provide a holistic view of the social, economic, and environmental status of the country and its states and UTs. To date, there is no other measure available that shows the performance of states and ranks them accordingly. This is sufficient to boost energy among states to outperform and achieve the targets by 2030. It has been designed to provide an aggregate assessment of the performance of all Indian states and UTs. It will help government, policymakers, businesses, civil society, and the general public to understand the key areas in which respective states/UTs need to invest and improve to fulfil the set targets.

The basic tenet of governance relies upon the pillar of data. Good governance can never be practiced without proper availability of information. A lack of data impairs governing bodies in managing even basic affairs in a target-oriented mode. Therefore, constructing the SDG India Index by the government is one of the path-breaking steps in a world where "Data is the new oil" (Humby, 2006).

In eight years, the Indian government has successfully embarked on the journey of the SDGs. If it continues at this pace, by 2030, India will be among the top countries in achieving the SDGs, and it will also be Viksit Bharat.

BIBLIOGRAPHY

Diaz-Sarachaga, Jose Manuel, Daniel Jato-Espino, and Daniel Castro-Fresno. "Is the Sustainable Development Goals (SDG) Index an Adequate Framework to Measure the Progress of the 2030 Agenda?" *Sustainable Development* 26, no. 6 (2018): 663–671.

Ghosh, Nilanjan, Roshan Saha, and Soumya Bhowmick. "SDG Index and Ease of Doing Business in India: A Sub-National Study." *Occasional Papers* 199 (2019).

Jain, Garima, Aromar Revi, J. Koduganti, and A. Abbas. *Localising SDGs for India*. Indian Institute of Human Settlements, Bengaluru (2018).

Schmidt-Traub, Guido, Christian Kroll, Katerina Teksoz, David Durand-Delacre, and Jeffrey D. Sachs. "National Baselines for the Sustainable Development Goals Assessed in the SDG Index and Dashboards." *Nature Geoscience* 10, no. 8 (2017): 547–555.

3 Janata + Data
Making Data Science and Participatory Planning Come Alive

Poornima Dore, Paresh Manohar, and Shruti Yerramilli

ABSTRACT

Effective local governance is key to development that is inclusive and sustainable, at the local level as well as the national. Data, evaluation, learning, technology, and analysis (DELTA) is a unique framework developed by us for a participatory approach in microplanning so that the development plans truly reflect the needs of the beneficiary community. The framework encompasses the collection and analysis of datasets for institutions (schools, anganwadi, village, gram panchayat, health centers), households, and individuals and has been carried out in several geographies. The DELTA Pilot in Chandrapur (Maharashtra) was conducted by the Tata Trusts in partnership with the District Administration of Chandrapur, Maharashtra, in three of its blocks, including 115 gram panchayats. Subsequently, an intensification program was initiated in the three blocks focused on the implementation of the village development plan prepared based on this participatory DELTA process. The list of activities planned for the year were painted on a common wall in the village, along with their status. The intent has been to encourage people to review the current situation of their village development and actively leverage local institutions for timely fund flow and action. We share some of the experiences and learning as a use case of implementing this framework.

Box 3.1: Chandrapur was one of the first districts where data, evaluation, learning, technology and analysis (DELTA) was rolled out. The district comes under the Nagpur division of Maharashtra. Nearly 33% of the total land area of the district is under forest cover, and 43% is under agriculture. It has large reserves of coal and limestone and is thus a major center for mining and industries in the state. The total population of the district is 21,94,262 (Census of India, 2011). There are 15 blocks in Chandrapur with a total rural population of 14,24,424 (Census of India, 2011).

A lack of real-time data about the institutions and households in the village and the low reliability of secondary data sources in the departments made it a challenging task for the district administration in Chandrapur to identify the areas which lagged in infrastructure availability and service delivery.

3.1 INTRODUCTION

"Effective local governance is key to development that is inclusive and sustainable, at the local level as well as the national."[1] In a diverse country like India, this is the only way to ensure that the development narrative at the center is not the representation of the average or majority but an integrated

DOI: 10.1201/9781003487531-5

approach that reflects the collective needs and aspirations of the people through a decentralized approach.

Historically, the government of India (GoI) has favored a planned approach to development since independence. This was done with subsequent five-year plans, which were "centralized and integrated national economic programs."[2] Steps towards bottom-up planning are quite recent. While the 73rd and 74th amendments gave local governments constitutional recognition, the 14th Finance Commission's step to assign funds for them through state governments was one of the first steps towards empowering local governments. Conterminously, the Ministry of Panchayati Raj initiated the Sansad Adarsh Gram Yojana, which recognized gram panchayats as the unit for implementation of development policies through the creation of village development plans (VDPs).[3]

However, none of these endeavors outlined methods that would effectively involve the community in the planning process, to make them stakeholders in their development plan. Unsurprisingly, the Goa Institute of Public Administration and Rural Development's training manual for preparation of VDPs also lamented that "people's participation in the planning process remains a distant reality."[4] The other constraint was the lack of capacity in the local governments to fully own the process of drafting the village development plan and its implementation. A crucial bottleneck was the lack of data that could supplement this process.

A lack of reliable and updated data led to inefficient fund allocation and lack of convergence among government departments. As a result, the plans have not been inclusive enough to meet the aspirations of the bottom of the pyramid, and it led to staggered development in villages.

Decentralization also requires people to give direct feedback to the concerned authorities and vice versa. However, historically, there was no mechanism for decision-makers to connect with people directly. While there are formal grievance redressal channels, they are ineffective in the absence of adequate information at the disposal of the community members.

For the most part, governments on various levels have responded to the need for better data management through the digitization of records and MIS. However, data mainly was in silos, not "talking" to each other and often showing conflicting pictures, thereby raising issues of reliability and mutual distrust. Furthermore, there are often limitations in the ability to utilize this data.

Thus, while these large data sets of public records existed with the government, their use was constrained by how compartmentalized the information was and how infrequently most datasets were updated. For example, our decadal census process based on an exhaustive survey exercise that informs our government often becomes ineffective in delivering granular efficiency in the implementation of schemes and policies built on the base of data sets nearly a decade old.

While large data sets have been used effectively by companies to drive their top line and bottom line, they have not been effectively utilized for planning by governments. There have been huge administrative challenges to reduce gaps in the planning process from national, state, to district blocks and ultimately at village levels. Recognizing this, all major national initiatives and missions like the Sarva Shiksha Abhiyan, National Health Mission, National Rural Livelihood Mission, and National Rural Employment Guarantee Scheme have strategized implementation focusing on decentralized planning.

Data, evaluation, learning, technology and analysis (DELTA) as a process and tool was conceptualized by us and taken up as a Tata Trusts initiative to strengthen data-driven governance in conjunction with communities. This involves bridging existing gaps and supporting the community-driven village development planning process through technology and community mobilization capacity building. It is envisaged that this institutional strengthening process and informed, data-backed decision-making will enable villagers to have ownership of the planning process and find solutions to larger issues that plague their communities.

DELTA has been successfully implemented in partnership with the district administration across five states in India: Vijayawada (Andhra Pradesh); Chandrapur (Maharashtra); Jajpur, Kendujhar and Balasore (Odisha); East Singhbhum, West Singhbhum, Saraikela, and Noamundi (Jharkhand); and Kolar (Karnataka).

We will now elaborate on the DELTA experience (with reference to Chandrapur district in Maharashtra) from pilot initiation to its scale-up by the government of Maharashtra to the entire district for microplanning based on the framework.

3.1.1 DELTA

DELTA is a framework for a participatory approach in microplanning so that development plans genuinely reflect the needs of the beneficiary community. The framework encompasses collecting and analyzing datasets for institutions (schools, anganwadi, village, gram panchayat, health centers), households, and individuals. While micro-planning approaches to development have been tried before, they have never been tried on such a massive scale or with a completely digitized survey collection process.

Saansad Adarsh Gram Yojana (SAGY) propelled the creation of DELTA, which was designed in a way that kept various stakeholders (community members, electoral representatives, members of local administration, civil society organization, pillar institutions like schools, hospitals, anganwadis, etc.) at the center of the process. Once the stakeholders are mapped, the DELTA process takes a four-pronged approach to support community-driven development planning.

Box 3.2: The DELTA Pilot in Chandrapur (Maharashtra) was conducted in three of its blocks in 2015–16, including 115 gram panchayats. Subsequently, an intensification program was initiated in the three blocks focused on implementation of the village development plan prepared based on the primary survey. It also focuses on inculcating the habit of regular gram sabhas in the region and making everyone aware about the progress made so far. The list of activities planned for the year is painted on a common wall in the village, along with their current status, and is regularly updated. The intent is to encourage people to analyze the current situation of their village development and seek accountability from elected representatives and government officials. For district administration, the pilot provided a clear picture of infrastructure and service delivery status in the region. Thus, they could redirect the resources in the regions which needed it most. Encouraged by the results of the pilot, the district administration decided to scale it up in the entire district.

- First is the identification of all stakeholders of the planning process in the geography. This means that the on-boarding of stakeholders like government officials, citizen forums, electoral representatives, and civil society organizations is indispensable to rolling out a sustainable planning exercise that is fully owned by its beneficiaries and principal actors. The most important part is the role and contribution of the political and administrative leadership. In this case, the local champions were members of the legislative assembly and the district collector. Their buy-in and advocacy has been essential in this.
- Second is the detailed survey process. To collect data, volunteers from the beneficiary communities are intensively trained, creating community resources. These volunteers become the backbone of the community mobilization endeavors, which is at the heart of ensuring an effective bottom-up planning process. The survey process uses the DELTA app to collect data concerning households (also individual-level data) and institutions within a village/hamlet (taken as the base survey unit).
- Third is the community mobilization process, where DELTA field experts work with the volunteers to sensitize community members about the status of their villages/settlements and mobilize gram sabhas to shape the village development plan actively. Participatory rural appraisal exercises provided both a rich source of qualitative information and essential tools to engage community members in defining the aspirations of the community and the constraints they face.

Box 3.3: A step beyond the pilot process, a district-wide primary survey was conducted by the District Administration of Chandrapur in association with Tata Trusts during 2017–18. The objective was to:

1. Collect data at the village level regarding demography; human resource availability; infrastructure and service availability; status of water and sanitation; and availability of basic facilities for household levels – electricity, water, toilet and cooking gas connection, road connectivity, and so on.
2. Develop a system to have near real-time information about the status of infrastructure, service delivery, and requirements at the village level, supported by evidence as applicable. It was to be developed on the lines of the gram panchayat develop planning (GPDP) process.
3. Use this system in prioritizing the requirements of people and high-need areas in planning and resource allocation.

- Fourth, and a crucial aspect of the DELTA process, is technical training of governance functionaries and members of the local administration. This is important to ensure that huge volumes of granular data are generated that help governance systems to migrate from departmental and siloed decision-making.

The activities under DELTA are conducted as a pilot in a region with active participation from government/local administration. Once familiarized with the approach, the local administration can then scale up the project. The data thus collected is hosted on a dashboard that allows for first-level descriptive analysis of the data and data visualization. We will discuss the dashboard in the next section as we touch on the different types of data collected through DELTA and the processes behind them.

3.1.2 Data Collection and Quality

The DELTA process involves primary data collection through three sources:

- First is an app/tab-based baseline survey covering 100% of institutions and individuals in a village. The survey instrument is a structured questionnaire. To collect a comprehensive picture of the village, the survey instrument includes questions covering every aspect of the community, including demography, economy, social structures, health, and sanitation. The institutional questionnaires cover a range of questions to assess the status of the institution: its efficiency, reach, resources, and functionality. The questionnaire was designed to incorporate the socio-spatial diversity of our survey sites.

Box 3.4: In Chandrapur, data was collected from schools, anganwadi centers, sub-health centers, primary health centers, and gram panchayats of the entire district. In total 1524 villages, 813 gram panchayats, 2448 anganwadi centers, 1465 schools, 293 sub-centers, and 54 primary health centers were covered in the survey.

- Second, participatory rural appraisal (PRA) exercises included transact walks, focused group discussions, and resource mapping. PRA exercises are interactive and build a sense of confidence between our field teams and members of the community. Often, they also become occasions for the community to self-reflect, and a lot of issues that they may want to address can be identified during these exercises. Such methods often indicate collective needs of the village/habitations.

- Third, using institutional records for better decision-making, for example, in the case of recording the number of cases of pregnancies in a given period and subsequent institutional deliveries.

The DELTA app allows these three types of datasets to be put together on one dashboard, which means, for example, we can not only see how many individuals reported a case of malaria in the recall period (found through survey data), but we can also find how many such cases were recorded at the PHC (institutional information) and may be able to recognize one of the main hurdles in early treatment like non-availability of ASHA workers in the concerned village/hamlet (information obtained through PRA exercises).

In all this, maintaining data quality is of the utmost importance. DELTA follows carefully designed technical checks and balances to ensure data quality, including automation of data error reporting built into the survey tool to send out data error flags are requisite intervals and maintaining a trail of errors reported to continuously calibrate the survey tool to rectify issues or support field teams as required. A fixed frequency of data error feedback to the surveyor also ensures timely support and refresher training, as needed. The data is also checked against identified benchmarks for every variable at regular intervals. For instance, given the reported productivity of a land in an area, the average produce from a unit of land can be used as a benchmark against which reported crop production can be checked. Extreme outliers, if any, can be identified, and data for them can be cross-checked. While these and other necessary technical safeguards are built into the process, we also want to talk about the human aspect of data.

A very important part of maintaining data quality is building adequate trust and awareness in the community. Some of these errors often occur because of issues like a stakeholder's lack of trust in the process leading to withholding or misinformation. For example, information on ownership of land or income is often inaccurate owing to trust issues. Similarly, aspects like health, adequate training and sensitization of the volunteers, and deeper interaction and awareness campaigns for community members are required.

Throughout the DELTA exercise, an open channel of communication is maintained with the community. All relevant information – like eligibility to access schemes, status of the village and progress along agreed-upon targets like the number of toilets built against those required – are shared with the stakeholders through wall paintings at resource centers in panchayat or other community buildings and awareness campaigns. This open channel ensures people actively engage with the process and follow its progress. The coordinators also maintain detailed action plans to monitor progress in the achievement of targets.

3.1.3 Cornerstones of a DELTA Exercise

Creating community resource persons – The purpose of the DELTA process is not to plug into the existing governance mechanisms to support bottom-up development. Rather, it aims to build technical and human resource capacity, starting at the grassroots level, with the aim of handing over a sustainable process to the rightful stakeholders (community, elected members, and local administration) at the end of the project period.

Together with our field implementation partners, we undertake intensive training for our volunteers, not only for the survey and PRA exercises but also on the entire process of development planning. Our field teams are structured in a three-tiered system with district coordinators, bloc coordinators, and village coordinators, reaching out to the last individual in the communities.

The coordinators work closely with elected members and governance functionaries to bridge the information gap between them and the community members. The volunteers are also important points of contact for people and provide handholding support to the community in implementation of priority items in the plan and access to the services for individuals. Box 3.5 elaborates on one such success story.

Stakeholder-ownership of the process – Perhaps the biggest achievement of the DELTA process has been to generate a feeling of ownership among the community members with respect to the data and the planning process. Information dissemination in the community about the progress through wall paintings and public discussions is an important part of the process. Our field experiences in Chandrapur showed that involving stakeholders in the process resulted in proactive participation on their end. Being able to track the progress of a certain project that the members collectively agreed upon builds trust and encourages them to voice their concerns and aspirations and learn to build it in the framework of the planning process.

Box 3.5: Delta success story. The process of compiling village development plans through the DELTA microplanning process, along with implementation partner Janseva Gramin Vikas Va Shikshan Prathisthan, in the Mul, Pombhurna, and Jiwati blocs of Chandrapur district began in March 2017. The data that was collected showed 18 households that did not have access to LPG connections, even though they were eligible beneficiaries under the Ujjala scheme. It was also found that the households were not aware of the scheme and its benefits. Our village coordinator, Mr. Salman Pathan, educated the households about their rights and also worked with the cluster and gram panchayat coordinators to reach out to the concerned bloc office. The coordinators assisted in submission of the required documents on behalf of the beneficiaries and also followed up with the bloc office. By December 2018, all eligible beneficiaries had been covered under the scheme.

Unlike most surveys, as part of the DELTA exercise, the results of the survey and PRA exercises are discussed with the stakeholders. This also acts as a round of quality check. In collecting information from many respondents, there is always a chance of errors seeping into the data. As the results are discussed with the stakeholders, they may point out any discrepancies, which can be crossed-checked and verified.

3.2 THE DASHBOARD

The data collected through the DELTA exercise is hosted on a dashboard and can be viewed for different variables along different geographies (village/bloc/district). The dashboard also holds the compiled village development plan. It is a digital medium that helps visualize and spatialize data points – for example, a chart could summarize the ratio of girls to boys attending primary schools in different villages in a bloc and a map can show the location of these schools across the bloc). Figure 3.1 shows our dashboard.

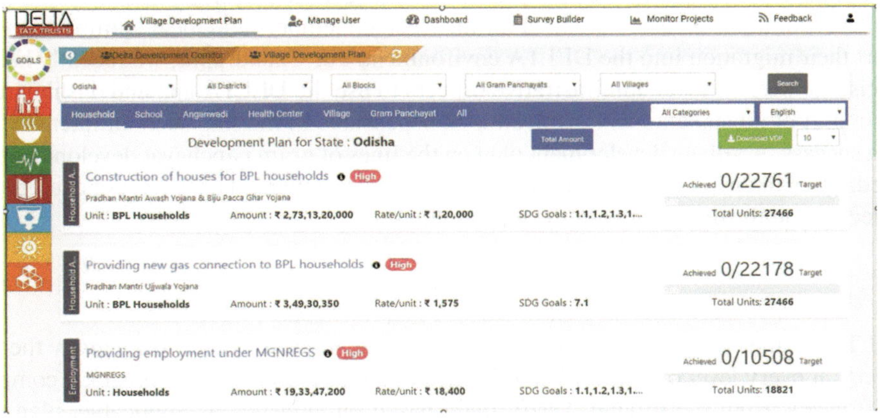

FIGURE 3.1 Sample DELTA Dashboard.

The final output of the survey is a near real-time dashboard. The complete data set can be viewed through summary charts across variables and geographies (village, bloc, and district), with periodic data update features. For instance, the following snapshots show the number of cases across villages in Odisha that have reported a lack of anganwadi infrastructure facilities. The chart shows the number of cases, whereas the map allows us to view their location (the map is interactive, and more granular information can be seen as we zoom in). To complete the picture, a detailed tabular list of the same variable (in this case non-availability of infrastructure facilities) can also be viewed, with a list of such centers/locations, along with evidence (e.g. images captured for toilets, buildings in poor condition, duration since electricity has been out of order). The dashboard also hosts the village development plan on the lines of GPDP, which includes the anganwadi plan, school plan, health center plan, and village plan.

The dashboard also hosts the digitized PRA records for every village. This makes it a powerful tool to take a holistic picture of a village and acknowledge the competing needs or interrelated issues of the community, paving the way for regional and social equity in an informed planning process.

Box 3.6: In Chandrapur, a total of 416 gram sabha were organized during the project period, on August 15, January 26, and October 2 and per the requirements of villages in the Mul, Jiwati, and Pombhurna bloc. A total of 68,892 community members actively participated in the gram sabha and discussed the 20 priority items in their VDP, selection of beneficiaries among them for government schemes and benefits, and review of the status of plan implementation. A total of 396 women's gram sabha have been organized before the main gram sabha in the Mul, Jiwati, and Pombhurna bloc, with total participation of 46,482 women for discussion on women's issues and the village development plan.

In the three blocs of Chandrapur, Mul, Pombhurna, and Jiwati, there have been 24,646 individuals who directly received scheme benefits, documents, and so on during the period April 2017 to March 2019. Around 50,500 households have been indirectly impacted.

3.3 BRINGING THE PLANNING PROCESS HOME

Finally, we discuss the crux of the issue – an effective, community-driven village development plan. All efforts of data collection and statutory requirements of creating a VDP are ineffective if the stakeholders remain alienated from the process. It was important for the stakeholders to treat the DELTA process as an initiation into evidence-based planning process and not as a one-off exercise.

As elaborated upon in Box 3.6, community mobilization in participatory planning, that is, encouraging people to hold the gram sabha and track implementation progress regularly, was a key outcome of the exercise. It is important for community members to understand what collective aspirations and issues look like and the information sets behind them.

Technical training exercises were held for the elected members and local administrative officers to support their migration into the DELTA environment. For Chandrapur, we recently held a training exercise for over 90 local administrative officers to use the DELTA app and dashboard.

The support of all stakeholders involved is indispensable in making such an intensive planning process a success. A village development plan on the lines of gram panchayat development plan was developed, after intensive discussions with community members. The VDP depends both on the data collected and the discussions in the gram sabhas.

3.3.1 Advantages of the DELTA System

The DELTA system (including the dashboard and the process framework) gives the planning department, at every level of administration (including district, bloc, and village), a complete suite to identify needs from the ground, verify them based on evidence as applicable, plan activities, allocate budgets (the Plan Plus feature of Panchayat ERP Suite being promoted by MoRD (GoI)), monitor the implementation using Action Plus, and verify the implementation against the real-time data reported in the system and social audits.

The DELTA system also allows for near real-time reporting/update of data. The survey app and/or data update interface can be used to regularly update data. This is supported by a facility to check historical data. For example, did the anganwadi have toilets by the due date of August 21 last year? This feature also allows tracking of data updates, including when and by whom. This ensures transparency and deters tampering with data.

Post-data update, the DELTA environment becomes a single source for village- and institution-level data (gram panchayat, anganwadi, health center, and school), breaking down the traditional departmental data recording system. This also means that once the data is fed into the system, the platform can be used to automate plan preparation (no manual intervention required).

3.3.2 Way Forward

DELTA is a sure-footed first step towards creating a responsible, data-driven planning/governance culture.

In the spirit of decentralized planning, one of the biggest steps forward would be the routine utilization of frameworks and platforms like DELTA for updating data by assigning responsible persons or office holders in villages, panchayats, anganwadi, health centers, and schools to report data using the app. It would be important to formalize the data generation process (status/requirement would be reported) from the point of the community (bottom-up).

For the public at large, we need to move towards opening such systems to public access and provide a mechanism for citizens to raise grievances regarding the status reported in the system; for example, if a school has it reported in the system that it has an electricity connection in each classroom, but the local stakeholders find this wrong/incorrect, then they can raise a grievance regarding it. This will ensure a built-in system of social audit.

NOTES

1 UNDP. "Local Governance and Local Development". Last Accessed on 23 April 2019. https://www.undp.org/content/undp/en/home/ourwork/democratic-governance-and-peacebuilding/responsive-and-accountable-institutions/local-governance-and-local-development/
2 MOSPI. n.d. *Chapter 7, Five Year Plans, Statistical Year Book of India.* Ministry of Statistics and Programme Implementation, Government of India. http://mospi.nic.in/sites/default/files/Statistical_year_book_india_chapters/Five%20Year%20Plan%20writeup_0.pdf
3 Saansad Adarsh Gram Yojana. http://saanjhi.gov.in/aboutus.aspx
4 Goa Institute of Public Administration and Rural Development. "Training Manual for Master Trainers on Preparation of Village Development Plan under Fourteenth Finance Commission, 26th–30th October 2015".

BIBLIOGRAPHY

Chambers, Robert. "Participatory Rural Appraisal (PRA): Analysis of Experience." *World Development* 22, no. 9 (1994): 1253–1268.

Chandramouli, C., and Registrar General. "Census of India 2011." *Provisional Population Totals*. New Delhi: Government of India (2011): 409–413.

Narayanasamy, Nammmalvar. *Participatory Rural Appraisal: Principles, Methods and Application*. Sage Publications Ltd (2009).

4 Using Big Data to Create an Evidence-Based Design for a Gram Panchayat Development Plan to Encourage Targeted Development

Rwitwika Bhattacharya, Aawanti Singh, and Aparajita Singh

ABSTRACT

A massive data collection exercise was carried out across schools, houses, panchayats, and health centers in 49 gram panchayats of Balasore, Odisha, to create development plans for each village to encourage targeted development through community governance. These data points were then boarded onto a dashboard, and a priority matrix was created across sectors, based on which village development plans were framed.

Critical insights were generated upon data analysis: 18% of villagers did not hold an Aadhar Card, and only 8.2% of households used LPG as the primary energy source for cooking. Village officials then carried out targeted interventions to alleviate the villagers' problems, and the previously underused gram panchayat funds were put into these data-backed actionable tasks.

4.1 INTRODUCTION

Two seemingly independent yet parallel developments within the field of governance have led to the need for local governments to adopt evidence-based policymaking. We can see the first one as growth in the power and status of panchayati raj institutions (PRIs) through both legislation and policy changes. The second type of trend is an increasing usage and perhaps reliance on data for policy design and implementation, which has been taking place worldwide.

Within India, panchayati raj institutions have been established as the third tier of governance, after the central and state levels. With the 14th Finance Commission (FFC), gram panchayats are entitled to INR 2 lakh crore from the central state revenue share. In parallel, data-driven policy design and implementation have become a widely accepted policy paradigm. International policy bodies like the World Bank have dedicated verticals to data-backed governance. Within India, the central and some state governments have created dashboards to evaluate and monitor the most prominent government programs like MGNREGA and Swachh Bharat Mission. These dashboards are the recipients and presenters of live data, enabling accountability and better planning.

4.2 CURRENT SCENARIO

In essence, gram panchayats have become increasingly empowered politically, administratively, and economically with the FFC. At the same time, data has been accepted as a critical policy design

DOI: 10.1201/9781003487531-6

and implementation tool. However, there is no convergence between these trends, which becomes especially problematic, as under the guidelines of the FFC, the creation of a gram panchayat development plan (GPDP) has been specially recommended. Data usage has not percolated to the third tier of the government, that is, local governance bodies.

As they are increasingly given more powers and responsibility, they need to institute evidence-based design within all developmental programs, as this is an accepted practice that has led to results at higher levels of government worldwide. The Balasore Micro Planning or Lok Yojana Project aimed to service this very need through collecting data from the panchayats which would be utilized for making development plans for each village under the panchayat. The purpose of the project was to encourage targeted development through community governance.

4.2.1 Why Odisha?

The location of the micro-planning project was chosen with much thought. The data collection exercise was conducted in all 49 gram panchayats (GPs) of two blocks, Balaiapal and Basta, of the Balasore parliamentary constituency in Odisha. The state enjoys a good record of decentralized planning and rural administration through initiatives such as the proposed Anchal Sasan Model of Governance in the 1950s. Anchal is a local authority considered one level higher than the gram panchayat. Post the 73rd Amendment, the state government was one of the first in India to organize panchayat elections. Based on these developments and the general political ecosystem in the region, the roots of local governance were solid in the state.

4.2.2 Potential and Challenges for Gram Panchayats

With the strong support of the state administration to local governments in the state, the GPs in the Balasore constituency were expected to embody the essence of responsible and accountable governance with a strong focus on community demands. However, ground reports, interviews, and discussions revealed that the GPs lacked any direction about the planning process because the plans that emanated from the GPs did not cater to the true developmental requirements of the region.

The lack of targeted development also led to massive unutilized funding, amounting to almost 50% of funds allocated to the panchayats. The GP members did not have the tools required for efficient resource planning management. About 50% of the panchayats did not create an annual GPDP, and out of the ones that did, about 30% of them were last year's plans rehashed, with demand added for an additional structure or two. The need for change, improvement, and capacity building was realized by the office of the Member of Parliament, Balasore, which contacted the Swaniti Initiative, a data-based policy implementation organization in New Delhi.

4.3 SOLUTION

The Swaniti Initiative, in partnership with the Member of Parliament's (MP) office, conceptualized the Balasore Micro Planning or Lok Yojana project with funding support from Tata Trusts, one of

Box 4.1: Specifications of the survey conducted.

Survey details
- 2 blocks
- 6 types of surveys
- 49 gram panchayats
- 400 surveyors
- 580 villages

the biggest philanthropies in India. The project was executed with the aid of the ground organization Bharat Gyan Vigyan Samiti (BGVS), which consists of a national volunteer group spanning 23 states. Through consultations with the MP's office, a developmental plan of the project was created. The project entailed several components: (i) collection of data through six separate surveys, as showcased in Box 4.1; (ii) creating a dashboard for the collected data; (iii) data insights from the dashboard would lead to a priority matrix of sectors, for which targeted interventions could be designed; and (iv) based on the dashboard, individual village development plans (VDPs) were made constituting the GPDP.

The dashboard was a particularly crucial component, as both the VDPs and targeted interventions in the priority sector were planned based on data insights of the dashboard. The intervention design was carried out in consultation with the block officers to ensure and encourage buy-in from the local authorities. As far as the VDPs were concerned, data insights constituted the major portion of their construction. What complemented the quantitative insights were the qualitative exercises in the form of focus group discussions (FGDs) and interviews with gram sabha members who contributed to the VDP. The process was long and complicated, spanning a total period of eight months from December 2015 to July 2016, while the focus of the project was to enable the local authorities to collect data and empower the GP members to utilize this data for VDPs.

4.3.1 What Happened on the Ground?

Since data collection was the first step of the process, the project inevitably began with questionnaire creation. There were six types of questionnaires created for different demographics and facilities: (i) village level, (ii) household level, (iii) health facility, (iv) anganwadi center level, (v) panchayat survey, and (vi) school survey. The reason the survey exercise was pan-sectoral is because the purpose of the survey was twofold.

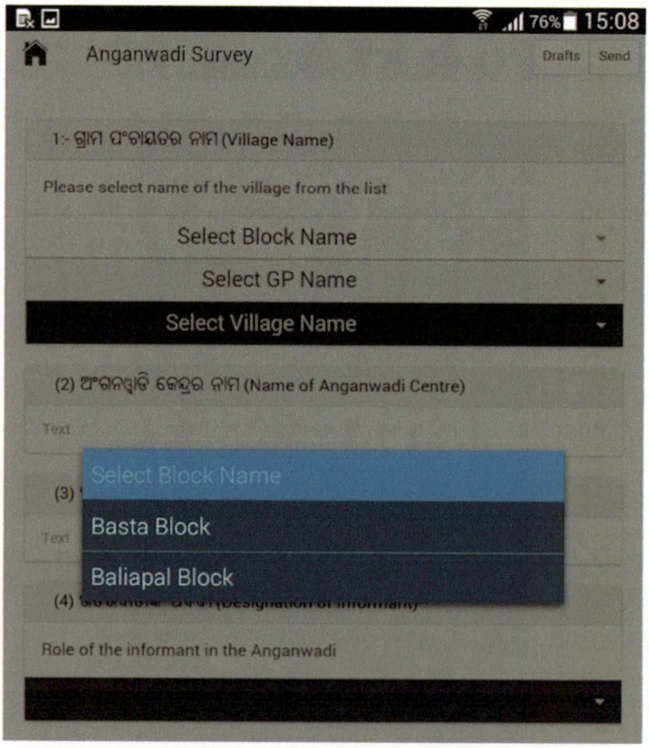

FIGURE 4.1 In-house survey application.

First, to catalyze micro planning, it was essential to get an idea of all developmental aspects of a region with a focus on health and education. Second, to increase the roles and responsibilities of the local government system, it was essential to assess the robustness of PRIs.

4.3.2 Volunteer Training

The training was the most exciting part of the data collection process. It was a tall order, with 400 volunteers from BVGS to be trained for a CAPI survey, as it was an in-house survey application, which the technology team had created. The volunteers were part of the national network and from the constituency. From people who had never used anything more than a radio to those who were adept with smartphones – the challenge lay in bridging the knowledge gap and ensuring they were all at a basic competency level before the data collection process began.

The surveys in the app were divided into major sections. This helped gauge the existing infrastructure and fund utilization of each of the 580 villages across 49 gram panchayats. The knowledge gap among the volunteers and the lack of stable internet connection in the interiors helped stress-test the application and strengthen the questionnaire coverage. There were several funny and genuine questions, such as, "Sir, how can we see that the data cloud has reached your office in Delhi?" or "What should I choose as an occupation for a beggar or pensioner?" which helped to distill the logic of any question.

While many of these questions could be answered, a number of them warranted discussion during the training session, which led to the inclusion of new questions in the survey form and yet more sleepless nights for the technology team. A leading example is, "Sir, for families who had ration cards together and then split up but still divided their monthly ration – which household should the ration card belong to, and how do we account for subsidized rations availed by both the families?" These questions made us realize the iterative process of questionnaire design, and some of the questions were changed based on volunteer training and the pilot survey. Some examples of change included removing the color of ration cards and having pensioners as a category in occupation.

With much discussion, the volunteer training portion of the program was completed. The beginning of the actual data collection entailed day-to-day monitoring and frequent quick-fix solutions to the lack of internet. The "offline sync" feature on the in-house survey application had to be developed through data collection. Some delays were endured due to the lack of connectivity; however,

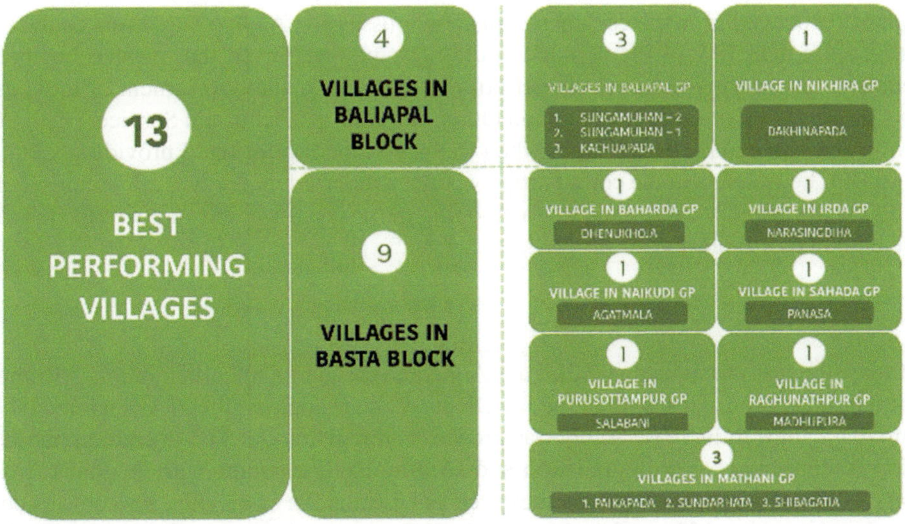

FIGURE 4.2 Sample dashboard of the best-performing villages.

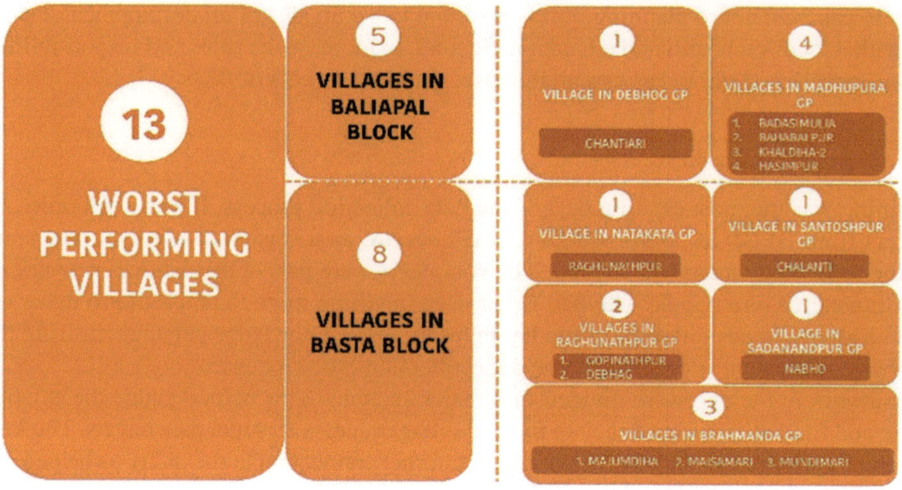

FIGURE 4.3 Sample dashboard of the worst-performing villages.

the data collection process was completed within a month through constant improvement in the survey application. Data on facilities and households based on differing metrics, like electricity access and waste management, was collected. Geospatial analysis was used to map the performance of different villages across varied dashboard metrics.

4.3.3 Capacity Building

Post dashboard creation, there was training for local authorities on how best to utilize the dashboard to design sector-specific interventions. The members of the gram sabha were also familiarized with how to use the dashboard so that they could formulate holistic VDPs. Through coordination with the district collector's and MP's office, internet was installed in the panchayat bhawan so that members of the gram sabha could access the dashboard. In the first few months, the Swaniti team analyzed the data on the dashboard to suggest designs of what a targeted intervention would look like.

It was essential to pilot a project based on the collected data and designed through a community participatory process. For instance, when data about possessing Aadhar cards was analyzed, it was found that more than 18% of the population did not own one. Given that Aadhar has been made a prerequisite for carrying out direct benefit transfer (DBT) by the government of India under the Aadhar (Targeted Delivery of Financial and Other Subsidies, Benefits and Services) Bill, 2016, this was a worrying situation. Furthermore, when the government prioritized providing LPG connection to all BPL households in the country, the two blocs reported that only 8.2% of households were using LPG as the primary energy source for cooking purposes. A sample output of the prioritization exercise is shown in Table 4.1.

4.3.4 Adhikar Abhiyan

Accordingly, the team designed an Adhikaar Abhiyaan campaign in collaboration with the district and bloc administration to strengthen the Jan Dhan–Aadhar–cooking fuel (JAC) trinity. It consisted of organizing camps to increase Aadhar awareness and registration. The overall emphasis was on providing Aadhar cards to the population and seeding bank accounts with Aadhar cards to take advantage of direct benefit transfer schemes such as Ujjwala. An orientation session was conducted by the assistant program officer of the blocs with all the sarpanches, who passed on the information to the anganwadi workers and ward members.

TABLE 4.1
Sample Prioritization Exercise

Prioritized Issues (Issues as Validated by Data/Statistics)	Action Points	Line Departments to Leverage	Budget Estimate
144 families demanded work under MGNREGA last year (2015), of which 120 families worked for an average of 45.77 days per family.	At least 980 persons, including 469 skilled wage laborers and 511 unemployed persons, must submit written work demands under MGNREGA.	MGNREGA (Block & GP)	N.A.
Inadequate availability of MGNREGA work.	At least 150 days of MGNREGA work will be available to 980 persons.	MGNREGA (Block & GP)	N.A.
Need for better earning opportunities through livestock.	1,025 livestock-owning households can be targeted for increasing their earning through training and equipping them with better livestock management tools.	Animal Husbandry Department	N.A.
739 households are landless.	739 households must be provided with homestead or agriculture lands.	Revenue Department	N.A.

Using this governance channel, information was passed on regarding the Aadhikar Abhiyan details. The local stakeholders were brought together to streamline the operational details of the campaign, be it venue or schedule, to create a collective onus among the authorities. This led to an essential insight about data-driven programs, that intervention design based on the sharpest data insights requires robust buy-in from local authorities for sustainability.

4.4 IMPACT

During the implementation, certain shortcomings in the entire registration process for Aadhar cards were brought to the fore. Some issues, like the lack of a proper format and increasing public awareness through the camps, were handled through the Adhikar Abhiyan. However, specific issues required more attention. During the campaigns, it became evident that there was no fixed rate for changing the information specified in an individual's Aadhar card. Address changes and misspelled names were the most common changes required, yet no policy mechanism catered to these needs.

This required policy-level changes from the state authorities, and accordingly, the matter was moved up to the district collector (DC). The DC forwarded the request to the chief secretary. Finally, the government of Odisha released an official notification specifying the rates applicable across the entire state for updating or changing information on an Aadhar card. This was a perfect example of how evidence had led to higher-level policy changes due to ground-level demands emanating from data insights.

This is only one example of a data-backed intervention. Data was collected about various sectors ranging from anganwadi centers to the number of teachers available per government school to the frequency of Village Health Sanitation Nutrition Committee (VHSNC) meetings. All these data points were utilized for the creation of individual village plans. The data dashboards built led to the formulation of several VDPs, which contributed to 49 GPDPs.

5 Enhancing Reader Engagement and Creating Interaction through Data Analysis

Sugata Srinivasaraju

ABSTRACT

Big data was analysed and sifted to present an interactive experience to readers during the Karnataka elections of 2013. Data was incisively sliced to help governance and the government's welfare programs. All the data relating to candidates, including their wealth and qualifications, was collected and then presented in the form of colourful roulettes, cartograms, treemap visualisations, and bubble charts that provided critical information to the reader entertainingly and insightfully. The exercise received an overwhelming response from the public.

5.1 INTRODUCTION

The 2013 Karnataka Assembly poll was approaching. The only thing playing on my mind was that we should be innovative in our coverage. The standard things like analyses, interviews, surveys, and ground reports would happen in the newspaper. That was the routine. The quest was to do that one big thing: putting our coverage a few notches above our competitors. What that one thing could be, we weren't clear until then!

I had crossed over from an English newsroom (*Outlook* news magazine) in 2012 to editorially lead *Vijaya Karnataka* (VK), a Times Group publication. Therefore, 'competitors' did not mean the narrow sliver of Kannada newspapers but also included English dailies. VK was the largest-selling news product in the southern Indian state of Karnataka. Its circulation was around 6.5 lakh at the time, and the daily average readership was close to 4 million. Given the profile of the paper's readers, which was distributed across various urban and rural tiers, all ideas of experimentation were tied to an unstated awareness that it should engage a cross-section of readers, if not most of them.

During the 2004 and 2008 Karnataka Assembly elections, along with two election-junkie friends, I had studied the 224 assembly constituencies in Karnataka minutely. I had looked at caste numbers that were available with political parties; used data on the Election Commission site to analyse the behaviour of each constituency until then; pulled out average margins of victory in previous elections; and took into cognisance demographic shifts after delimitation of constituencies. To all this, we had added our insight on the existing political and social climate. Everything was fed into a roulette in our heads, and we made an amateurish prediction of the number of seats each party would win. It was a stick-your-neck-out survey. The 2008 research we even published on a site to a mild consternation in political circles.

5.2 CURRENT SCENARIO

There was a vague method, journalistic experience, and ground reports in what we had done. There was deductive reasoning and common sense, but not exactly data science or psephology that the

DOI: 10.1201/9781003487531-7

biggies in the field like Prannoy Roy or Dorab Sopariwala had made Indians familiar with. Our success rate the first time was over 70%. The second time, it improved, but not much. Accuracy apart, there was huge reaction and reader engagement.

5.2.1 Roulette to Cartograms

We wondered if this kind of an election roulette could be built for the newspaper. A step further, could we make it even more interactive by involving readers in the prediction of seats?

We knew our strengths. VK had a solid network across the state. Therefore, logistics for such an election game would not be difficult, I speculated. When this was passing through my head, I ran into Sajith Pai, a corporate colleague in the Times Group at a routine meeting. In the course of our conversation, he mentioned that there was a data science start-up called Gramener that was trying to interest the senior management of the Times in reading enormous data at its doorway in a somewhat easy and edible fashion.

This seemed like the solution we'd been looking for. I asked if they could create graphics with election-related data, as we weren't familiar with data visualisations and cartograms at that point. Pai said the simplest thing for me to do would be to have a chat with the start-up. Meetings were organised at lightning speed. Soon, one of the founders was in my office.

In his genial voice, one afternoon, he gave me a crash course in what the company was doing and how incisive slicing of data helped governance and the government's welfare programmes. He showed how cartograms helped spot and decipher social and behavioural trends and how it made the understanding of complex issues very accessible. I asked if it could be applied to election-related data, he said they had not thought in that direction so far but were curious to collaborate.

At VK, I put together a team of four people. It included one junior assistant editor who had shown some promise with data work, B K Ganesh, and a trainee sub-editor to assist him. A senior graphic designer and a junior would assist him. The first task before us was to identify themes to build cartograms and check the availability of data. We decided that we would dedicate the page opposite the edit page, which was free of ads, once a week to cartograms.

But before we took off with the project, there was a hitch that had to be sorted out. Since a special budget and vast space in the newspaper were involved (space in the Times' parlance is 'real estate' and is often measured with the same ferocity), I discussed the project with colleagues in the management. They were broadly appreciative of the idea but were not sure if so much space should be allotted for a data experiment.

At that stage, tentatively, the proposal was for eight full pages in the run-up to the polls. They raised some very valid questions, which were roughly as follows: We are not an English newspaper; would this grand exercise mean anything to our semi-rural and rural audiences? Will it sufficiently engage them? Will we be able to sell more copies if we simply announce the special pages on the front page? Does it require separate marketing? Can't we do with half pages rather than full pages? Why eight pages; can't we be more modest? Can we train our own people to do it rather than bring in an external collaborator?

My team and I had no specific replies to their queries, but we wanted to ensure that the project was not shelved. Without having worked out anything at all, and what could be read as bravado in retrospect, I said that these cartograms will be made interactive. Even a reader with high school education will be made to play with the visualisations we'll put out. For its novelty, it could generate more interest than a crossword, a Sudoku, or other stock game strips that run in a corner of the paper, I emphasised.

When I mentioned gamification, that suddenly created interest around the table. Thankfully, nobody at that point asked me to explain this further. On the contrary, they added to the idea by saying if interactivity and gamification were planned, a contest could be organised on days the special pages appeared, and readers could be given prizes. They even started debating possible sponsors. The final result of the negotiations was that we settled with five full pages of cartograms instead of eight.

5.3 SOLUTION – FIVE PAGES, FIVE THEMES, FIVE GAMES

The first page was published on 26 March 2013. It was based on affidavits submitted by candidates declaring their wealth. It was drawn from the 2008 polls, as the nomination process for the 2013 polls were still underway and the data was incomplete. The banner headline for the page was 'Karnatakada Kubera Nayakaru' (Kubera was a mythological reference, and it meant 'Kuberas of Karnataka Politics'). The straplines read this way: "What you see here is a recreation of the universe of our legislators' wealth' (the word used in Kannada for 'universe' was 'brahmanda.' In the local context it evoked a sarcastic tone, because the same word is idiomatically used in a phrase to suggest humongous corruption).

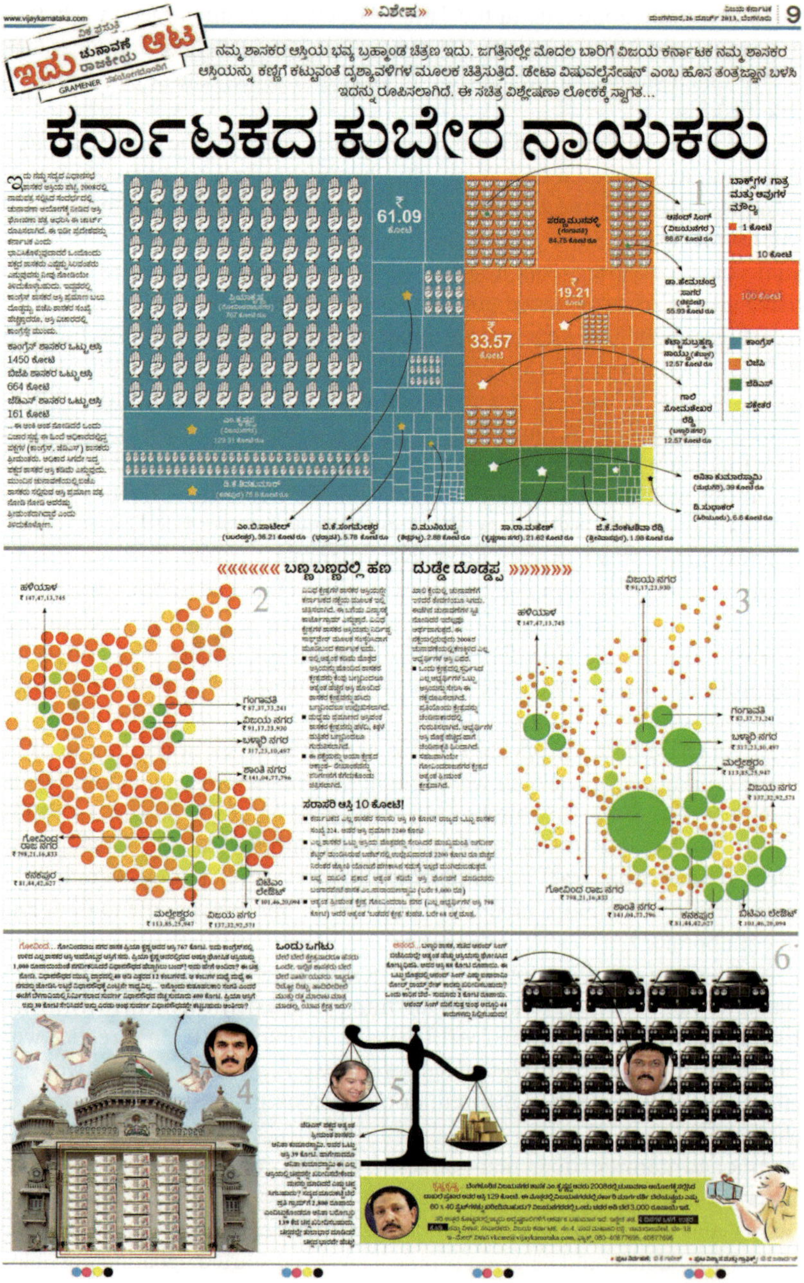
www.vijaykarnataka.com
» ವಿಶೇಷ »
9
ನಮ್ಮ ಶಾಸಕರ ಆಸ್ತಿಯ ಭವ್ಯ ಬ್ರಹ್ಮಾಂಡ ಚಿತ್ರಣ ಇದು. ಜಗತ್ತಿನಲ್ಲೇ ಮೊದಲ ಬಾರಿಗೆ ವಿಜಯ ಕರ್ನಾಟಕ ನಮ್ಮ ಶಾಸಕರ ಆಸ್ತಿಯನ್ನು ಕಣ್ಣಿಗೆ ಕಟ್ಟುವಂತೆ ದೃಶ್ಯಾವಳಿಗಳ ಮೂಲಕ ಚಿತ್ರಿಸುತ್ತಿದೆ. ಡೇಟಾ ವಿಷುವಲೈಸೇಷನ್ ಎಂಬ ಹೊಸ ತಂತ್ರಜ್ಞಾನ ಬಳಸಿ ಇದನ್ನು ರೂಪಿಸಲಾಗಿದೆ. ಈ ಸಚಿತ್ರ ವಿಶ್ಲೇಷಣಾ ಲೋಕಕ್ಕೆ ಸ್ವಾಗತ...

ಕರ್ನಾಟಕದ ಕುಬೇರ ನಾಯಕರು

FIGURE 5.1 Treemap visualisation showing the wealth of political parties and leaders.

For the first time in India, Vijaya Karnataka created evocative images that give the citizen a perfect idea of the size of a legislators' wealth. These images have been created using a new scientific method called data visualisation. Welcome to a refreshingly illustrated world of analysis.

The page was divided into three cartogram decks, and the bottom spread was the game section. The first cartogram (Figure 5.1, a treemap visualisation, looked at wealth and political parties. Within each party, the proportion of each MLA's wealth was marked. The ones who took more space had the maximum wealth. The second cartogram (Figure 5.2, a scatter plot, looked at all 224 constituencies across Karnataka to plot the wealth of leaders representing them. The third, a bubble

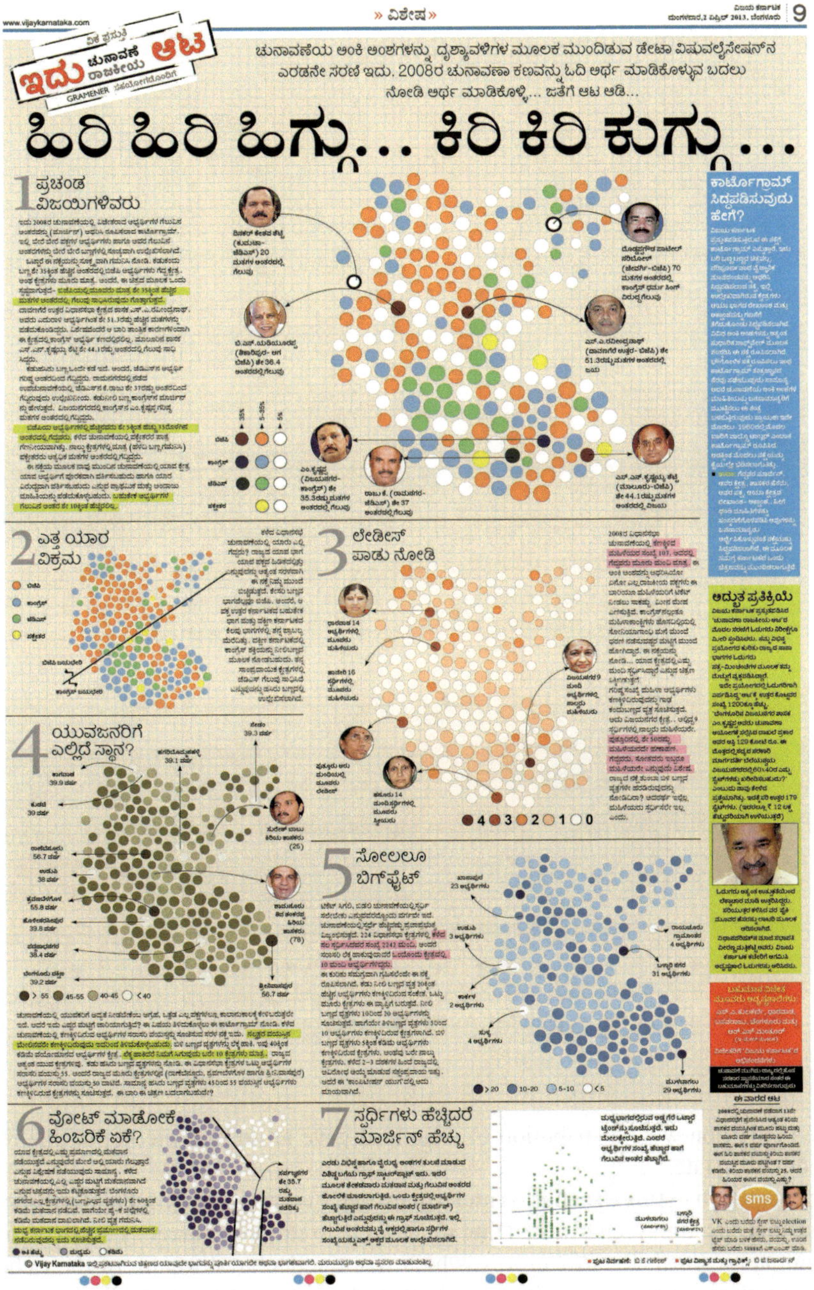

FIGURE 5.2 Election visualisation with victory margin, legislator age, and gender balance in assembly.

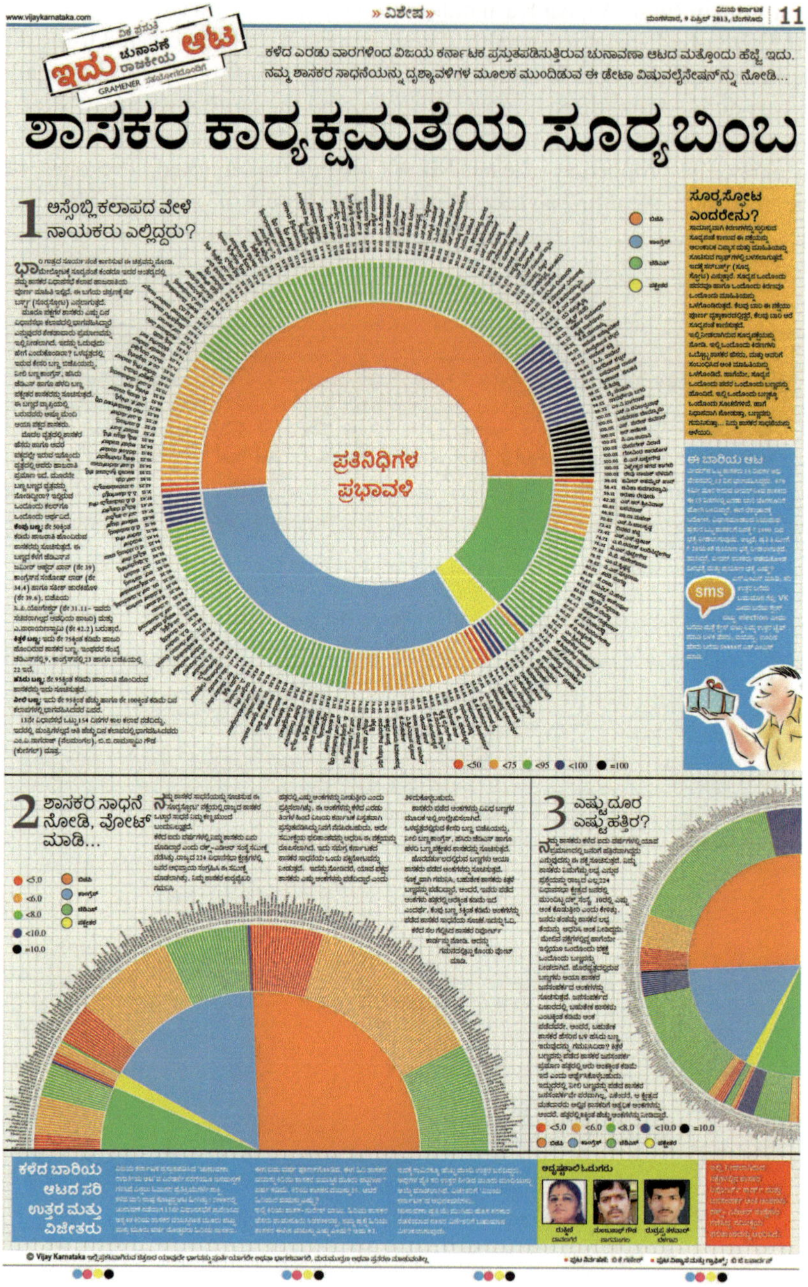

FIGURE 5.3 Performance of the elected MLAs based on citizen survey.

chart, looked at region-wise distribution of wealth. It gave an instant picture of where wealth was concentrated in the state.

The game section had three scenarios and a question. Figure 5.3 highlights the three scenarios in this visualisation. The first scenario said that if one were to stack up currency notes of one-thousand-rupee denomination worth Rs. 767 crores, which was the declared wealth of the richest Congress MLA, Priya Krishna, then that money would block the entrance to the Vidhana Soudha, the seat of power in Karnataka. The main door of the Vidhana Soudha was 40 feet high, and there were 12 pillars around it. If the currency notes were placed in between these pillars, then the entrance would be blocked. It was suggestive of what money power would potentially do to peoples' power.

The second illustration told the story of the kilograms of gold that could be purchased, at the rates prevalent then, with the declared assets of Anita Kumaraswamy, the richest MLA of Janata Dal Secular. The third illustration spoke about BJP Bellary MLA and mining baron Anand Singh. It said 44 Rolls Royce cars could be purchased with his declared wealth.

Given these three scenarios, the question to the reader was simple: The declared wealth of M Krishnappa, the Vijayanagar Congress MLA and father of Priya Krishna, was Rs. 139 crores How many 60-by-40–foot residential plots could be purchased in his constituency at the guidance value prescribed by the government? The burden was on the reader to determine the guidance value. M Krishnappa was also known as 'Layout Krishnappa' because he was a real estate tycoon and had created many residential layouts in and around Bangalore.

The second page was published on 2 April 2013 (Figure 5.2). There were six cartograms of bubble charts and scatter plots. They looked at margins of victory, age of legislators, and gender balance in the assembly. There was a game question here as well. The results of the first contest were also announced on this page. Since the entries for the first game question were so numerous, the winners were picked in a lottery by Veeranna Matigatti, the chairman of the upper house of Karnataka's legislature.

The third page was published on 9 April 2013 (Figure 5.3), and it was made up of marvellous-looking sunburst cartograms. They looked at the performance of MLAs. The data was from a survey that the newspaper had published. The survey graded each MLA on a scale of ten on nearly a dozen parameters.

The fourth page, published on 15 April 2013 (Figure 5.4), was made of bubble charts that looked at questions asked by legislators in the assembly and their attendance. A fascinating insight on this page stated how the kind of questions younger MLAs asked differed from those asked by older members. The split of subjects were apparent. If elders asked on traditional subjects like PWD, rural development, and social welfare, the younger generation hovered around medical education, industries, and exercise.

The fifth and final page, made up of bubble charts again, was published on 22 April 2013 (Figure 5.5). It looked at caste distribution in constituencies. There was also a chart that represented issues uppermost on voters' minds. The caste data for this page aggregated the figures available with political parties. The voters' data came from the newspaper's survey.

5.3.1 Issues During Development: To Jump Out of the Page

The pages looked aesthetically splendid, but there is a little story behind their creation and the panic it caused in the process. When I was introduced to cartograms, I had seen them on a computer screen. The representations there looked minute, colourful, and rich. At that point it did not occur to me that these cartograms had to be eventually transposed into print, in a much larger size, and newspaper design had its own demands.

The details and subtle colour gradients of the cartograms were impressive on a computer screen, but when the first models arrived and prints were taken, they looked crude. The gradients were too subtle, and the colours were getting smudged. Red, green, blue (RGB) colours were not working for a cyan, magenta, yellow, and key (CMYK) scheme. The overall aesthetics were unappealing. The cartograms were not jumping out of the page as we had expected them to. This led to the possibility of failure in print.

I sat with the senior graphic designer, B G Janardhan, and asked him if he had a solution. He said he could create a new graphic with every single cartogram produced. He would ensure they were to scale. The team would give us the cartograms, and we would recopy, enlarge, change colours to create contrast, use tiny pictures where necessary, sharpen edges, and bend everything to our design needs and standards. The light blue grid that we decided on as a backdrop for each page was inspired by my Moleskine squared notebook.

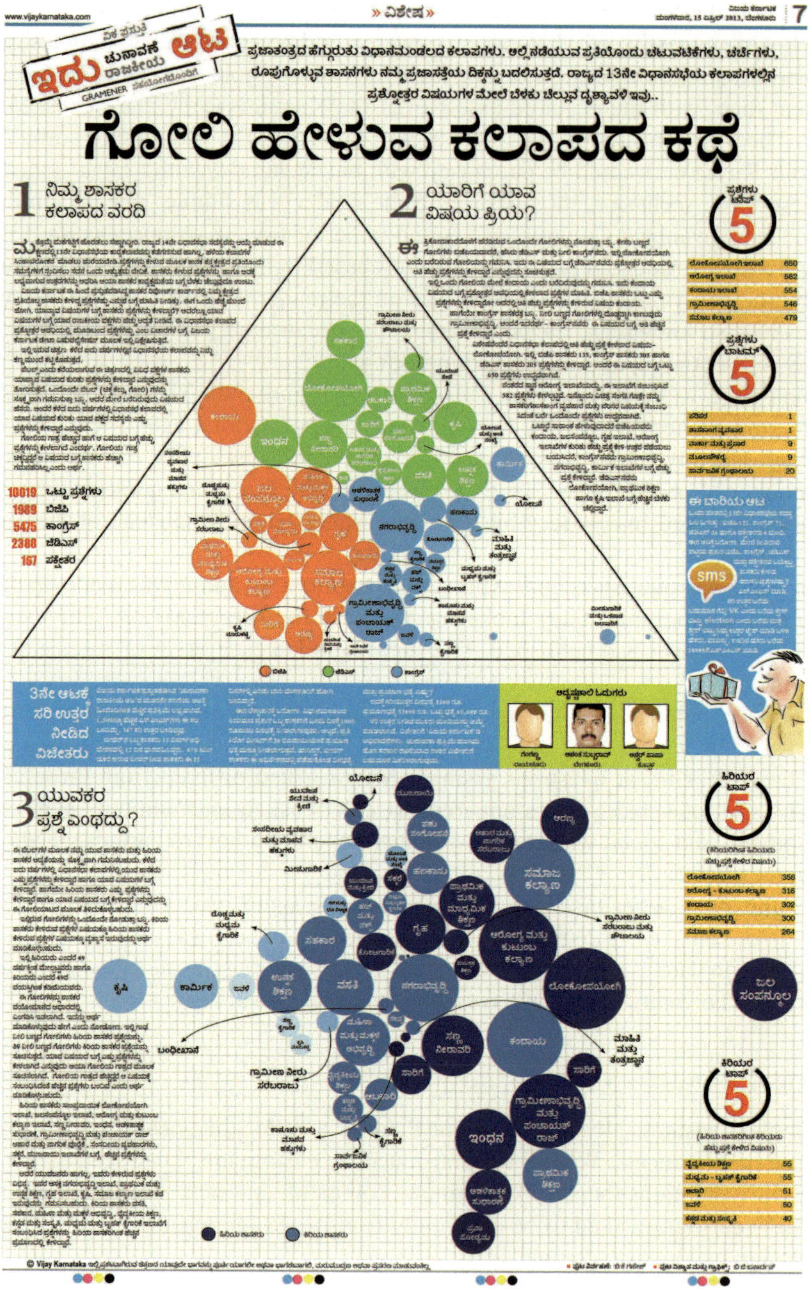

FIGURE 5.4 Analysis of questions asked by legislators in the assembly and their attendance.

The entire process was time consuming and intense, but that did not deter the designer. Our graphic designer asked us to reschedule the appearance dates of pages and give him some extra time. We reworked the schedule. He spent two to three days on each page sitting next to an engineer who was embedded in our newsroom for this rather funnily twisted exercise. The engineer needed to be there to guarantee the accuracy of data. Finally, the results on the pages made us beam and jump for joy.

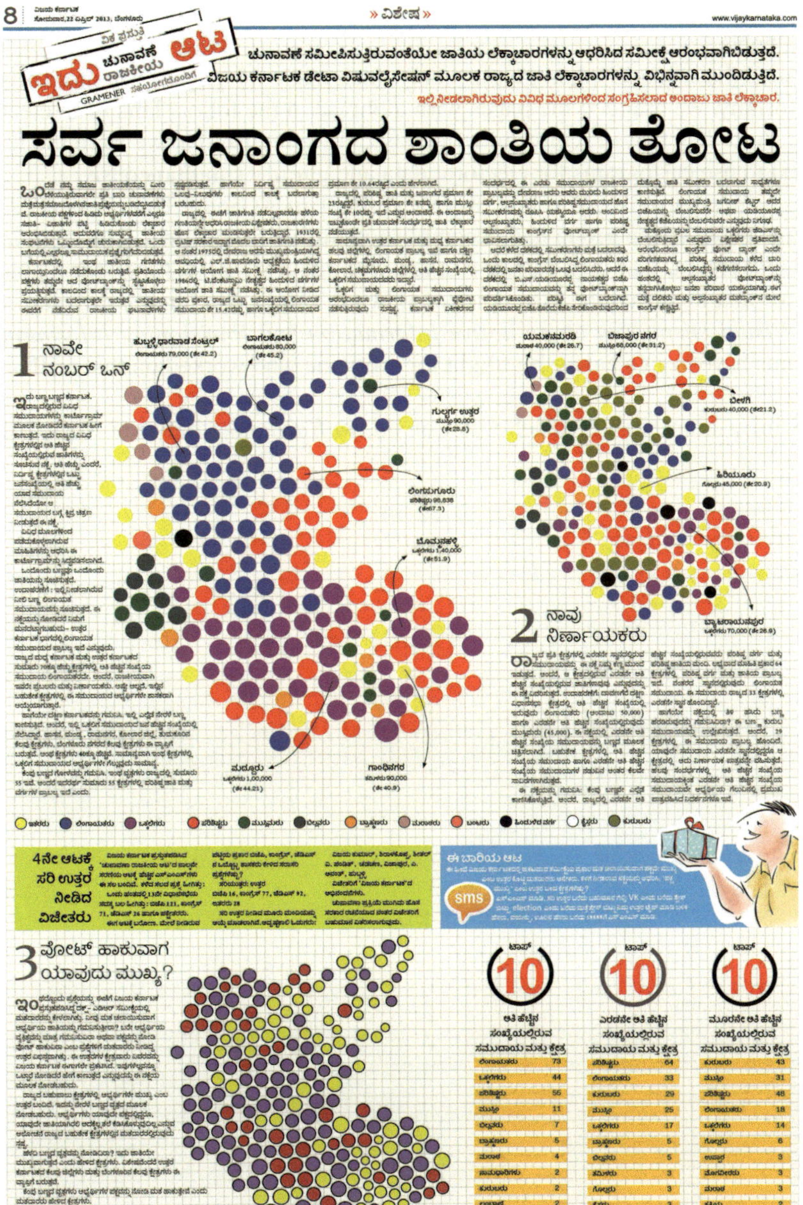

FIGURE 5.5 Caste distribution of the constituencies in Karnataka.

5.4 IMPACT

Soon, success was at our doorstep. The finest moments were when our network of correspondents and stringers started feeding us anecdotal evidence of the success of the cartogram pages – the tea shop chats, the lunch-break discussions, calls to local bureaus asking for clarification of the game question, attempts to find fault with our data. The decision to give prizes had added to reader engagement. The average time a reader spends on a page naturally increases. Of course, there was no definite measure of it as one would have with a digital product, but the yardstick was the number of email and SMS responses that reached us by 3 or 4 pm on the day the page appeared. The enthusiasm was palpable.

We published the winners' photographs, too, which made the pages even more attractive for the readers. Editorially, we branded the pages with a phrase in the folio: 'This is an election game, do play with us'. And they played. Only subsequently did we realise that we were the very first in the Indian context to use cartograms for electoral data and gamify simultaneously.

The pages were received very well across the entire Times Group. There was mild astonishment that a language paper of the group had accomplished this enormous feat. The technical director of the group, Sanat Hazra, who had earlier worked with the *New York Times* for a long stretch, was so appreciative that he gifted me his copies of Edward Tufte's two landmark volumes on data and design: *The Visual Display of Quantitative Information* and *Envisioning Information*. The election cartograms have been a career highlight for everyone who worked on that brilliant data-driven project.

(Sugata Srinivasaraju was the chief editor of *Vijaya Karnataka* between April 2012 and March 2015.)

6 Vidhan Sabha Proceeding Search – "Making Governance Transparent, in People's Language"

Anand Kulkarni and Raj Kumar Gupta

ABSTRACT

The Uttar Pradesh Legislative Assembly has preserved a plethora of documents recording the proceedings of the Legislative Assembly over the decades, primarily as physical copies. Earlier on, manual or physical searches were conducted in libraries to retrieve information by authorized personnel, taking a huge amount of time to retrieve information. Intelligent Search Mobile Application was developed based on the scanned, annotated, and indexed digitized assembly proceeding books, videos, newspaper clippings, and gazettes. This mobile app enables members of the Legislative Assembly (MLA) and staff to access the documents of Vidhan Sabha proceedings at their fingertips, making it efficient to find past information and perform official duties.

6.1 INTRODUCTION

Being the largest and oldest constituency of India, the Uttar Pradesh Legislative Assembly (UPLAS; also known as the Uttar Pradesh Vidhan Sabha, UPVS) has preserved a plethora of documents recording the proceedings of the Legislative Assembly over the decades. However, most of these documents are stored manually or as physical copies.

Earlier on, manual or physical searches were conducted in libraries to retrieve information. Authorized personnel would sift through the relevant shelves per the given sequence in which various documents were filed and placed. This manual method is imperfect, time consuming, and delays access to information.

Further, it needs more physical space to store massive amounts of documents that are bound to increase with time. Physical storage also requires constant maintenance over time, and documents need to be protected from damage from the elements. As such, a significant amount of capital goes into storing, maintaining, and protecting this data.

This meant interested parties had to retrieve information manually from age-old proceeding books, newspaper clippings, gazettes, and video clips. This is a challenge. First, it is a highly time-consuming task. If one needed to refer to a newspaper clipping for a particular proceeding, then an authorized person would have to visit the shelves containing the records and search between specific file folders, leaf through rims of papers, or visit specific libraries, and then retrieve the physical copy of the record of the proceeding. This copy would then have to be verified against the proceeding of the specific date and session. Therefore, the manual visit took an excessive amount of time for the information seeker and ate up human-hours of the authorized personnel retrieving such information.

DOI: 10.1201/9781003487531-8

Further, retrieving and associating data from various sources at different locations is also tedious. Also, since the data association was being done manually, there were instances of mismatching or gaps in information.

Against this backdrop, the Uttar Pradesh government aimed to digitize these documents so that they could be easily accessible and searchable. This would significantly benefit the relevant seekers of records of proceedings, researchers, and the public.

6.2 SOLUTION

The widespread adoption and use of mobile technologies have the potential to provide innovative ways to access, use, manage, and analyze information. The Uttar Pradesh government asked the team to develop a digital solution for the digitization and navigation of the Vidhan Sabha proceedings and related data.

A website was also created to access the proceeding documents, newspaper clippings, and gazettes. However, it acted more as a repository of proceedings, providing a linear way of searching between documents. For example, there was no way to search between the text of the various files available.

Another issue with the website is that once opened, all the related images, such as JavaScript (.js) and Cascading Style Sheets (.css) files, are loaded onto the webpage, making it data heavy and relatively slow to buffer and perform. On the other hand, the mobile app – the UPVS-Intelligent Search Solution – is much faster, as it only displays search request and response data on the screen.

Therefore, the existing methods were tedious as well as slow.

This smart solution platform is known as the UPVS-Intelligent Search Mobile Application. It has scanned, annotated, and indexed all the digitized assembly proceeding books, videos, newspaper clippings, and gazettes.

This mobile application (app) enables members of the Legislative Assembly (MLAs) and staff members of the Uttar Pradesh Vidhan Sabha to easily access the documents of Vidhan Sabha proceedings at their fingertips. The mobile application also assists these users with readily available information to help them perform their official duties in more efficient ways.

Further, the UPVS-Intelligent Search application also enables various other stakeholders to gain access to the information of the UPVS proceedings. The mobile application is available for download on both Android and iOS platforms.

6.2.1 Challenges during Digitizing of Documents

Creating a digital solution for colossal amounts of data is a challenge. However, there were some specific issues that the implementation team and the various stakeholders at UPVS faced and overcame during the development and implementation of the UPVS-Intelligent Search solution.

First, it was the digitization of documents. Turning roomfuls of physical documents into digital documents is a challenge. The documents contained information about the actual proceedings and summary, newspaper reports, gazettes, and so on. The first step, scanning this massive heap of documents, was the first real challenge.

Our next challenge was optical character recognition (OCR). During data collection, it was found that a lot of documents were available only as photographs or images. It then became vital that OCR be enabled on these documents to make them completely searchable. This was a separate challenge, but these documents contained valuable information and were being pulled from across various databases for the purpose of creating a complete and accurate database.

Another definite challenge turned out to be retrieving accurate information from age-old documents that were found to be torn, illegible, and in damaged condition. Some documents that were found in different languages were then translated by the team. At this stage, it was of the utmost importance that the meaning and substance of the text remained intact and nothing was lost in translation.

Another specific challenge was to collate documents stored within multiple databases located at several locations and convert them to a singular one-stop platform. Data was being transferred between various stakeholders and parties. Therefore, managing IP connections between all interested parties was a challenge.

It is known that a key technical element of data sharing is to maintain smooth data flow between the various stakeholders carrying, sharing, and exchanging data. The data flow must be undertaken so that it keeps the valid interests of each stakeholder intact. This often means following the chain of authority/hierarchy between relevant departments to obtain necessary permissions.

Media player integration with the mobile app was also a challenge. Usually, a media player is a separate platform, and the app is a separate platform, and integrating them within each other can bring certain challenges.

We wanted to provide quick access to users, so developing a mobile app was the best solution. This way, users could quickly access the records through their mobile devices. The app was created for both Android and iOS platforms.

It's important that the app have a user-friendly interface, especially browsing and navigating through search queries and search results, so that more and more people can benefit from this data.

Moreover, we specifically wanted to give easy access to the proceeding documents to Vidhan Sabha members and their staff. The app would be of the most assistance to them because it would provide them with readily available information which they can regularly use to help perform their duties more efficiently.

Also, the app must be easy to download, install, and operate at lightning-fast speed. It must provide bilingual support in both Hindi and English. We also made use of the common IONIC tool for devices with Android and iPhone platforms.

6.2.2 Key Features

Here are the key features of the app, as showcased in Figure 6.1.

Quick Access: The entire purpose of digitization of physical documents is to save time and provide ease to every stakeholder. Information seekers can download the app and get the relevant information on the go.

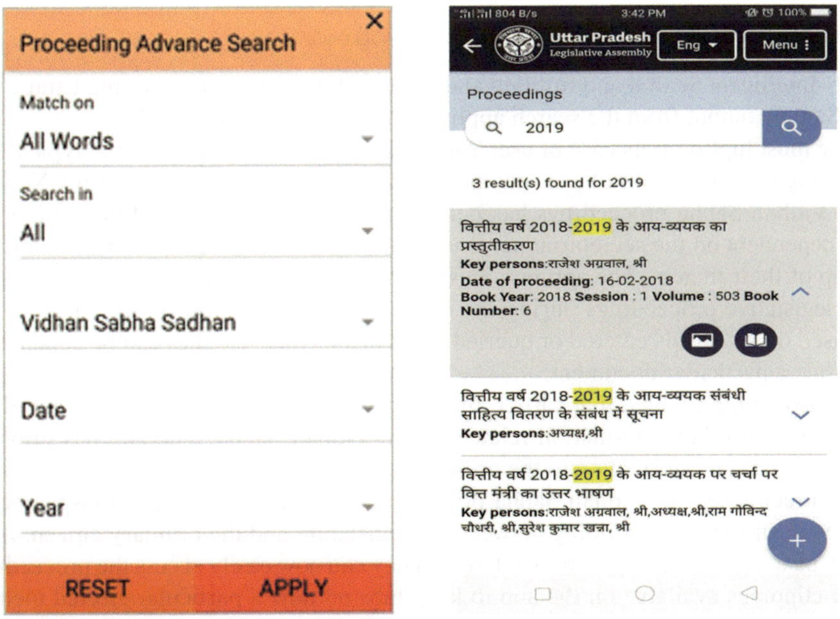

FIGURE 6.1 Key features of the search application for UP government.

Search Options: The app records data and provides a smart search option to retrieve specific information such as proceedings for a specific date or month or relating to a specific member.

User-Friendly: The app has a colorful and interesting user interface, keeping in mind visibility of features and ease of navigation.

Bilingual: The app supports both Hindi and English, so a language barrier does not obstruct users.

Indian Language Search Capability: The documents and the entire app, including all of its features, are available in Hindi. This means that users can search commands in Hindi as well. This is being done so that the ordinary public of Uttar Pradesh and the nation can also peruse the government proceedings and stay updated.

Custom-Built Scanning Software and OCR: A smart solution for digitizing documents does not only entail uploading scanned documents onto a digital portal. The UPVS-Intelligent Search app has kept the "intelligent" factor in mind by making these scanned copies readable through OCR. This means the scanned documents are converted into electronic text that can be highlighted, searched, and so on.

E-Book Creation: The app also allows users to collate information of the proceedings and ancillary information related to it into an e-book format.

Digital Vault: Privacy is paramount when collating, accessing, and using big data. The UPVS-Intelligent Search app stores data in a digital vault to keep the data safe and secure.

Data Centre and Disaster Recovery Centre: As noted, the privacy and sanctity of data collated and made available for access and use is critical. Keeping this in mind, we also created a dedicated data center that stores and manages all the data available on the app. Further, we also have a formidable disaster recovery system in case of data mishaps.

6.2.3 Challenges during Development of UPVS-Intelligent Search Application

The only criterion for developing a great app is to fulfill client expectations. However, expectations change as the blueprint of an app evolves. This was one of the significant challenges faced during the app's development stage. Ultimately, the team successfully finalized requirements per the request for proposal (RFP) scope.

6.2.4 Benefits of UPVS-Intelligent Search Solution

The UPVS-Intelligent Search app fulfilled and exceeded the expectations of the Uttar Pradesh government. Sample output from the search application is shown in Figure 6.2.

First, we must highlight its ease of use. Since this app was developed for the typical user who is not technically informed, ease of use becomes this app's first and foremost feature. Now, the accessibility of Vidhan Sabha proceedings has become convenient and quick. Information seekers are no longer dependent on the secretariat personnel, as they can retrieve all relevant information with a single tap of their fingers. The app provides seamless analytics capabilities for in-depth insights about the legislative proceedings during the house session. The app offers contextual search and content based on keywords entered or queried. This means that searches can be conducted even for the text inside a particular document.

It must also be highlighted that earlier, the common public did not have access to the proceedings, gazettes, and other information related to the Vidhan Sabha. The app has provided much-needed access to such information to the ordinary citizen.

Having direct access to Vidhan Sabha proceedings has brought greater transparency to public proceedings regarding trust-building between the legislature and the ordinary citizen. In a manner, the app can be used as a tally system for citizens. They can now easily access the proceedings or read newspaper clippings available on the app to keep tabs on how a particular elected member is performing or whether an elected member is performing their duties honestly or keeping their promises.

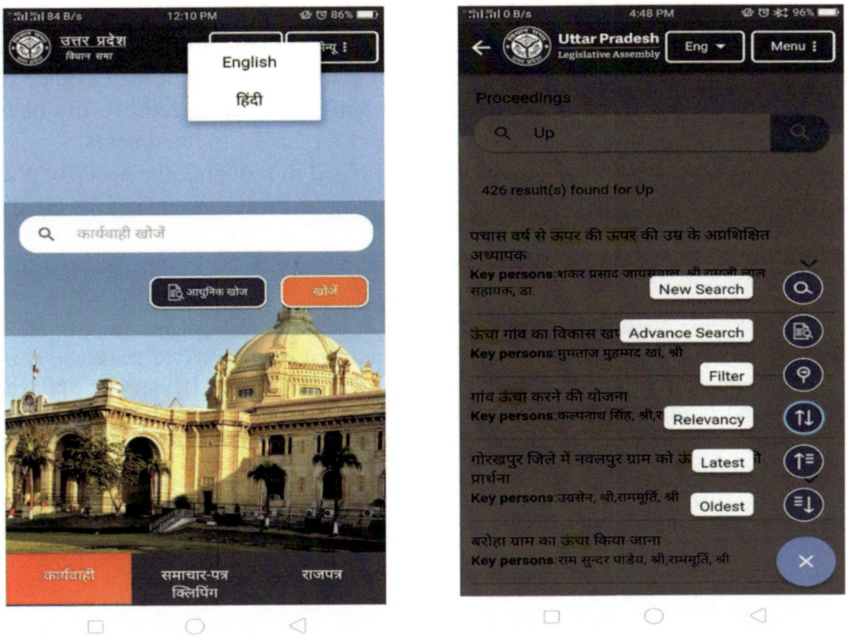

FIGURE 6.2 Sample output from the search application.

Further, the intelligent search feature of the app enables contextual search capability. For example, if a user searches the word "Yogi" within the app, the search results would display the word "Yogi" as a noun and an adjective. Also, it highlights occurrences of this word in other search results. Further, a user can also search for information relating to a specific date, word, source, year, and many different categories. For convenience, search results also provide several pages, words, and searches.

This app is a live example of how data being a non-rivalrous and non-excludable good can benefit the public. With this app, multiple users can now access the same document at the same time. Earlier, users would have to wait their turn either to access a document or make copies of the available records to be able to gain information. Now, multiple users can access the data without having to do so.

Another long-term benefit of the app is that because it provides all available information for a certain topic in one place, it can help users generate valuable data insights using cross-checking and comparing.

6.3 IMPACT

The general app users and Vidhan Sabha members and staff have given positive feedback, primarily over the ease of access created by the app. They prefer using the app because of its simplicity, smart search features, and quick turn-around. Since the app can be used by both Android and iOS users, it gives uniform and all-encompassing access to all users. Further, the app is not data heavy and thus doesn't require a lot of download time.

A key value indicator is that the app provides a single-point solution to access all types of data: search proceedings, newspaper clippings, and gazettes issued by the UPVS to both the common public and members. Further, special privileges of signing in have been provided to MLAs and other Vidhan Sabha members to view and download videos of proceedings.

The app's Advanced Search feature enables users to search for more specific information by applying filters. This helps users save time and energy. More than anything else, the app has made legislative proceedings available to the public anytime and anywhere in their choice of language. The dependency on a librarian or other authorized staff member to retrieve the documents has reduced to a bare minimum. There is no waiting period for access to documents.

Also, the app, being a purely digital solution, has helped in reducing the need for photostats and physical paper copies of the documents, thus contributing towards conserving the environment.

BIBLIOGRAPHY

Chinnakotla, Manoj Kumar, Sagar Ranadive, Om P. Damani, and Pushpak Bhattacharyya. "Hindi to English and Marathi to English cross language information retrieval evaluation." In *Workshop of the cross-language evaluation forum for European languages*, pp. 111–118. Berlin, Heidelberg: Springer, 2007.

Dolamic, Ljiljana, and Jacques Savoy. "Comparative study of indexing and search strategies for the Hindi, Marathi, and Bengali languages." *ACM Transactions on Asian Language Information Processing (TALIP)* 9, no. 3 (2010): 1–24.

Section II

Human Capital

7 Identifying and Rectifying Urban Bias in Allocation of Targets in Rural Skilling Program in Kerala

Harsh Nisar, Aakash Solanki, and Alka Upadhyaya

ABSTRACT

The Deen Dayal Upadhyay Gramin Kaushalya Yojna (DDUGKY) is one of the most important flagship schemes of the Ministry of Rural Development (MORD), Government of India. The main aim of the DDUGKY is to impart skills to the rural population, specifically youth in the 15 to 29 age group, in India so that they can earn their livelihood and live with dignity. This scheme operates in a public–private partnership (PPP) model where the MORD works with private skilling partners across various states to achieve India's skilling goals. Data analysis showed that the higher the proportion of a district's urban population, the more targets are allocated for the rural skilling program. To rectify this bias, an index was created using the Socio-Economic Caste Census 2011 and Census 2011. A district is rewarded if it caters to a higher number of deprived households and negative reinforcement for a district based on urbanisation. This index formed the basis for further allocation of youth for the rural skilling program. The implementation of the index in practice highlighted the role of socio-political considerations and practical tooling in the process of everyday data science within the public sector.

7.1 INTRODUCTION

The Deen Dayal Upadhyay Gramin Kaushalya Yojna (DDUGKY) is one of the most important flagship schemes of the Ministry of Rural Development (MORD). Launched in September 2014, the scheme aims to provide skill training and placement opportunities to rural youth from poor families, enabling them to secure sustainable employment and improve their livelihoods. The main aim of the DDUGKY is to impart skills to the rural population, specifically youth in the 15 to 29 age group, in India so that they can earn their livelihood and live with dignity. This scheme operates in a public–private partnership (PPP) model where the MORD works with private skilling partners (project implementation agencies; PIAs) across various states to achieve India's skilling goals.

To this intent and purpose, PIAs propose projects to the state along with estimated targets regarding various trades for which they would impart training to the rural population and the number of such candidates in each district that they propose to train. On a rolling basis, these projects are ratified by the nodal agency at the state and are cleared by an empowered committee.

7.2 CURRENT SCENARIO

While the ministry and its appointed PIAs were doing their part, the results were not as expected by the government. In some cases, candidates were easily mobilised for training, but they would drop

DOI: 10.1201/9781003487531-10

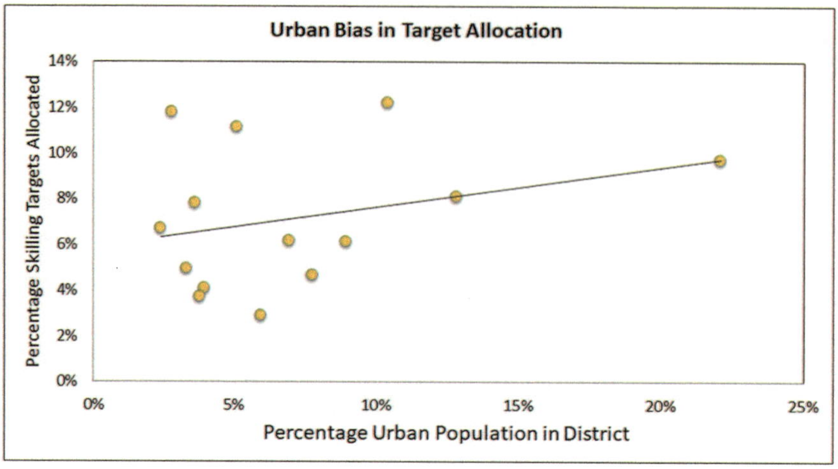

FIGURE 7.1 Urban bias in target allocation.

out during placements; in others, mobilisation of interested candidates itself faced challenges. Over time, the number of unfulfilled training and placement targets accumulated. As the targets for the program were limited, it was necessary that the right candidates for such a government intervention be targeted.

The authors, as part of the team situated at the Ministry of Rural Development and in collaboration with the State Rural Livelihood Mission (SRLM) of Kerala, inspected the data containing targets allocated across districts in the state of Kerala. We found a significant urban bias by the project-implementing organisations. Upon analysing the data, our team observed that the more urban population a district had, the more targets were allocated for the rural skilling program. This is the reason targets were not being completed: because the substantial population in the major areas targeted was the urban population and not the rural population. This trend can be seen in Figure 7.1.

The hypothesis was that PIAs, usually corporates and NGOs, were much more familiar with urbanised districts. It is also known that urban areas are more accessible and comfortable based on the location of their operations, vicinity, accessibility, ease of imparting skills, ease of setting up centres, and so on. It is also well known that there are usually inherent stereotypes around remote districts and the people living in usually inaccessible places, which can be seen with initial scepticism.

This, however, was not a good strategy, as it was against the intended objectives of the program, that is, to provide skills and thus greater employment opportunities to beneficiaries with lack of formal education and employment structures. On the contrary, persons in urban areas have much higher chances of having received basic education and vocational skills and are closer to employment centres, requiring less state intervention in seeking job opportunities. Therefore, the chances of finding individuals fit for the DDUGKY were higher in rural and remote areas, while more budget was being allotted for urban areas.

7.3 SOLUTION

The Strategy – Use of Existing Administrative Data for Planning and Targeting: The SRLM decided to correct part of the biases in the next lot of allocations and sought a quantified method of allocating targets. The target allocation was no longer dependent on the discretion of the PIAs but on data. With this intent, we created a "need" index using the Socio-Economic Caste Census 2011 and Census 2011 data to reward a district if it catered to a higher number of deprived households per SECC and used negative reinforcement for a district based on urbanisation per figures derived from Census 2011. In this manner, using the index, we were able to generate an ideal number of targets to be assigned to each district.

$$Gi = \frac{Ni * Ri}{\sum_{1}^{n} Ni * Ri}$$

where *Gi* is the percentage of total targets that should be allocated to district *i*, *Ni* is the number of households with three or more deprivations in district *i*, *Ri* is the proportion of rural habitations divided by total population of the district *i*, and *n* is the total number of districts in the state.

7.3.1 Development – Participatory Policy and Simplicity

One of the key factors was that the index that we prepared had to be simple enough to be understood, trusted, and accepted by the state administration and their officials who were going to take it forward with the PIAs. The state participated at each step of the entire process, and this is crucial because in the end, data alone can't replace the local context and experience of the state officials, and also, this approach leads to ownership of the outcome. Therefore, targets were constantly adjusted based on the inputs by the state officials, including political and administrative requirements (e.g. capacity of district officers to complete allocated targets, symbolism of allotting zero targets, etc.) which wouldn't have otherwise been captured by the index.

7.3.2 Implementation and Impact

For the fresh allocation of targets, each potential PIA had to be allocated multiple districts, and the numbers were to be optimised in such a manner that overall district allocation targets as decided by the index were maintained. As negotiation was expected from the PIAs to suit their preferences, a data-based system wouldn't work unless it could be updated in real time based on human inputs. Hence, an Excel-based tool was prepared (Figure 7.2), which allowed targets to be balanced across PIAs interactively during the pre-Empowered Committee meeting. The tool would allow the state officials to enter, add, alter, and amend district-wise targets for each PIA in a matrix. It would instantly update and highlight the total deficit/overflow in each district. All of this happened in real time, and therefore, decisions were held so that the final targets or allocations could still be achieved.

During the negotiation, if any PIA showed displeasure with the district targets it was being given, the state could adjust the targets elsewhere in the sheet and automatically update project- and district-wise deficits per the index. This made the entire process highly interactive and transparent.

The state officials effectively used the tool and allocated targets to impart skill to a population of more than 20,000 rural poor more equitably than previously. This is illustrated by Figure 7.3, where on the left-hand side, the Proposed Targets vs. Need Index is shown, and on the right-hand side, the Final Allocated Targets vs. Need Index is shown. Earlier, the Proposed Targets were scattered around the Need Index; however, after the implementation of the Index, the Final Allocated Targets are very much in line with the Need Index.

As a success story, the index was demonstrated nationally to all states. It came to light that similar biases were found at a national level. After implementing this index, the Ministry of Rural Development supported states to ratify and balance their targets. A policy for the purposeful allocation of targets across districts was drafted under DDUGKY and rolled out nationally.

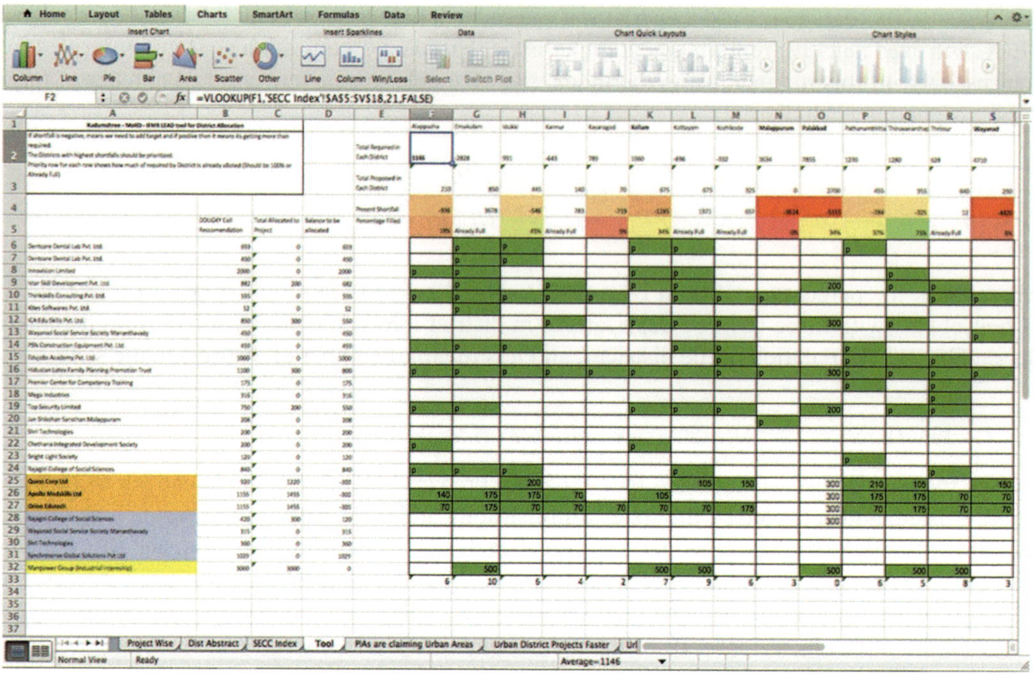

FIGURE 7.2 A screenshot of the Excel tool.

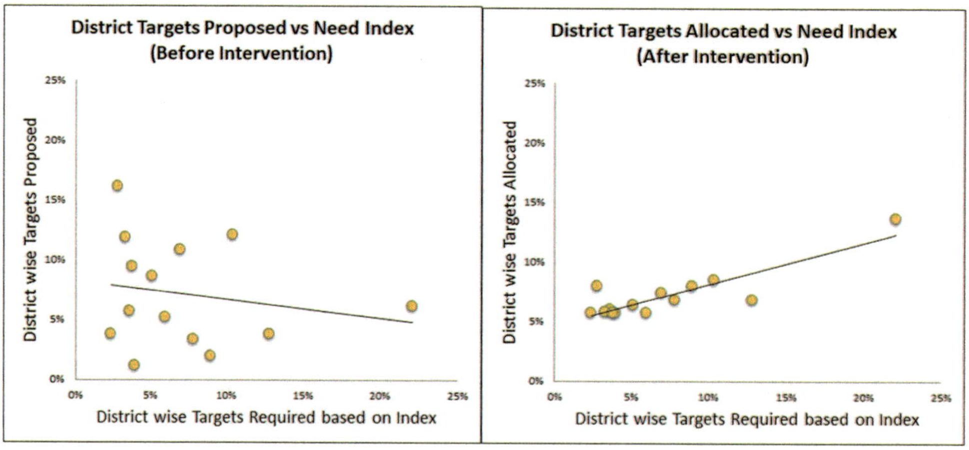

FIGURE 7.3 Comparison of results before and after the intervention.

BIBLIOGRAPHY

Cabral, Clement, and Rajib Lochan Dhar. "Skill development research in India: A systematic literature review and future research agenda." *Benchmarking: An International Journal* 26, no. 7 (2019): 2242–2266.

Elena, Ragazzi. "Effectiveness evaluation of training programmes for disadvantaged targets." *Procedia-Social and Behavioral Sciences* 141 (2014): 1239–1243.

Jamal, Tabassum, and Kasturi Mandal. "Skill development mission in vocational areas-mapping government initiatives." *Current Science* (2013): 590–595.

Mehrotra, Santosh, Ankita Gandhi, and Bimal K. Sahoo. "Estimating India's skill gap: On a realistic basis for 2022." *Economic and Political Weekly* (2013): 102–111.

8 Managing, Predicting, and Preventing the Coronavirus Outbreak in Orissa

A Data for Good Initiative

Kamaljeet Singh

ABSTRACT

The COVID-19 pandemic turned the world upside down. Systems and citizens struggled to cope with the massive effects of the coronavirus. The government of Orissa was in the same boat, struggling to provide healthcare to its residents and ensure health infrastructure in the state. For this purpose, technologies such as analytical dashboards and predictive modeling were brought to the forefront as they could solve problems that humans couldn't. Thus, the COVID dashboard came to life.

Data was collected from different healthcare departments, labs, hospitals, outpatient departments (OPDs), inpatient departments (IPDs), staff members, Indian Council of Medical Research (ICMR) forms, and so on. This data was then assessed and reviewed to develop evidence-based planning, monitoring, and supervision of the state population's health and real-time view of COVID-19 spread in the state. The system offered insight into disease outbreaks, prediction, and prevention within Orissa and was adopted by various other state governments.

8.1 INTRODUCTION

Indian state government healthcare ministries, respective agencies, and departments are primarily responsible for maintaining adequate health infrastructure for citizens in line with the central and World Health Organization (WHO) guidelines. The primary focus of the state ministry is to make provisions for hospitals, primary health centers (PHCs), healthcare staff (doctors, nurses, and support), ICU beds, pathology labs, and drugs and vaccines in proportion to the population of the state. In several states, this is managed through legacy applications, manual processes, unintegrated systems, and siloed data stores.

However, all this was put to the test in the year 2020, when the WHO announced the coronavirus (SARS-CoV-2) spread, a global outbreak infecting tens of thousands in more than 30 countries and causing the COVID-19 acute respiratory disease.

8.2 CURRENT SCENARIO

Such an event was unprecedented in the modern world, thus requiring governments (state and central) to take steps in attaining preventive, monitoring, curative, and rehabilitative direction. This gave a broader mission for the health department to contribute not only to the social and economic development of India by achieving the highest attainable high-quality health status and being accessible to the people of India but also being proactive in combating a disease outbreak scenario such as COVID-19.

DOI: 10.1201/9781003487531-11

States were swift to leverage the latest information technology for their health processes. The rapid deployment of processes to capture data and generate life-saving insights could help the government take proactive measures to manage the COVID-19 spread effectively based on these accurate data assessments.

In the first phase, governments were pressed with multiple challenges. First, it was difficult to track and trace infected individuals. Then, there was the huge task of building quarantine centers to cover all infected areas. Further, infected people had to be isolated from the public to contain the spread. The next challenge was to build a holistic capacity view, keeping in mind hospital beds, quarantine centers, clinics, and doctors across public and private healthcare facilities.

Other challenges before the government were managing the nationwide or statewide lockdown, controlling population movement and migration patterns, and monitoring disease spread. Finally, providing proper treatment overall and capturing treatment response was a massive challenge, as there was insufficient research and guidance about COVID-19.

Usually, data is retained in silos. For example, hospitals may collect information on patient status for their hospital, while the health department tracks and retains data on ambulance requirements in the city. But during an emergency like a pandemic, people can greatly benefit if all data is seamlessly available to the authorities simultaneously, so they can make critical decisions.

This required technology capabilities to consolidate the data/information flowing in from disparate sources, whether manual or application based, such as manually filled Indian Council of Medical Research (ICMR) forms and unintegrated hospital management systems, to one central application, which would provide a central analytics view to the respective governance bodies like municipal corporations, health departments, and hospitals. The state bodies needed to coordinate perfectly based on the spread, the availability of medical facilities, population movement, and so on. This, in turn, required accurate and timely information for planning, execution, and taking proactive measures.

8.3 SOLUTION

Our team worked with the government of Odisha to build a system of analytical dashboards and predictive models with advanced capabilities to collect data and clean and transform it with its Data Management product. SAS Viya's analytics further processed this to generate hidden patterns, key trends, anomalies, and predictions on the COVID-19 spread.

Our data and analytics technologies helped strengthen capacities for real-time data collection, assessment, and review for evidence-based planning, monitoring, and supervision for the health departments, presenting the view in real time on how COVID-19 was spreading fast in the respective state. The suitable platform processing datasets at volume, speed, and sophistication significantly impacted the ability to take more precise measures for containment.

The government was able to reduce the spread of COVID-19, initiate lockdowns in places with high infection rates, make provisions for medicines and equipment like oxygen concentrators, and present a real and timely view of information to the departments and the public through the public dashboard.[1] This made a huge difference in containing the spread of the pandemic and, ultimately, saving lives. The key analytical capabilities of the analysis and the dashboard are highlighted in Figure 8.1. These analytical capabilities were delivered with careful design and development, as seen in the following use case scenarios.

8.3.1 Data Integration across Healthcare Data

Much data was collected from different healthcare departments, entities, and documentation like labs, hospitals, outpatient departments (OPDs), inpatient departments (IPDs), equipment, staff members, and ICMR forms. However, no integrated data repository was available centrally for effectively tracking infected individuals or available capacities at various quarantine centers/hospitals at a central level, with details at a district to village level. Also, the data across different systems was non-standardized, which posed a great challenge in integrating this data.

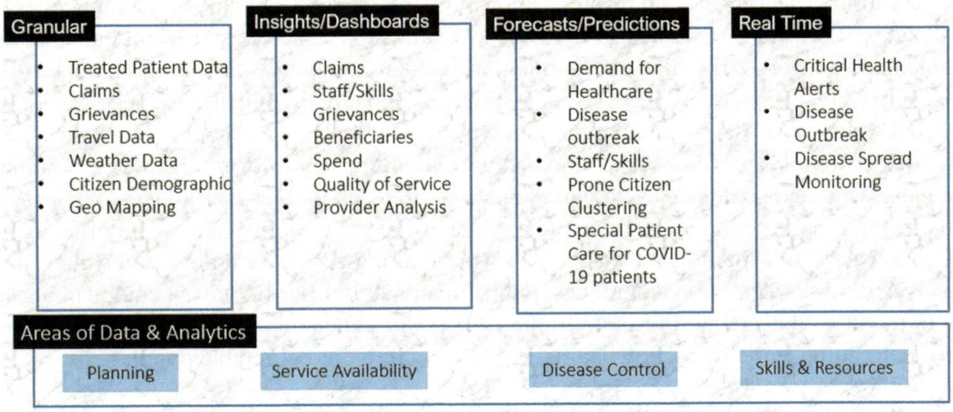

FIGURE 8.1 The analytical capabilities of the COVID-19 dashboard.

Our team helped the healthcare department integrate data from multiple silos, clean, and load the data into a central repository for effective planning, reporting, and decision-making. This helped the department in many data and analytics processes. First, they received cleaned and integrated data at a central location. All available data was presented in standardized demographic information such as city, state, and gender. Further, all the data available for analysis was reliable and accurate so that the health department could make impactful decisions. For example, integrating data led to effective management of resources, including rotation and allocation of available beds for COVID patients.

8.3.2 Singular View of Data

Once cleaned data was made available, the department could get a single view of the state across various aspects of spatial analysis (hospital and patient locations) and a time series (daily, weekly, and monthly) view on the spread, recovery, capacity utilized, and so on. This data helped the state government and the departments decide to impose or open lockdowns based on disease spread and migrant movements.

The department heads could make decisions based on a single view of the state. They could drill through, analyze, and identify more affected regions. They could also derive root causes and scenario analysis by linking hidden patterns. For example, one pattern that came to light was the impact of migrant inflow to various regions and the degree of lockdowns.

8.3.3 Disease Detection, Surveillance, and Vital Health Statistics

An automated system was built for monitoring COVID-19 disease outbreaks, which not only does baseline surveillance using advanced analytics but also leverages natural language processing (NLP) and machine learning capabilities to create models that process large amounts of data in real time and identify patterns. So, the model diagnoses a particular set of people with coronavirus through cluster analysis, pattern analysis, and segmentation.

Cluster analysis involves identifying groups of similar cases based on specific attributes. In the context of COVID-19, cluster analysis can help identify geographic areas, communities, or demographic groups experiencing higher rates of infection or transmission. By detecting clusters of cases, public health authorities can implement targeted interventions such as increased testing, contact tracing, and localized lockdowns to prevent further spread.

Pattern analysis involves identifying trends, patterns, and anomalies in data. For COVID-19 surveillance, pattern analysis can be used to track the trajectory of the pandemic, detect changes in transmission patterns over time, and identify emerging hotspots or areas of concern. Public health officials can make informed decisions about resource allocation, intervention strategies, and public messaging by analyzing patterns in case data, testing rates, hospitalizations, and other relevant indicators.

Segmentation involves dividing the population into groups based on shared characteristics or behaviors. In COVID-19 surveillance, segmentation can help tailor public health messages, interventions, and resources to different population segments. For example, segmentation based on age, occupation, socioeconomic status, or vaccination status can effectively inform targeted communication campaigns, vaccination strategies, and outreach efforts to reach high-risk or underserved groups.

Creating and maintaining a database and a system of alerts will give healthcare officials timely access to accurate information for critical decision-making. By leveraging cluster analysis, pattern analysis, and segmentation techniques, public health authorities in Odisha could enhance their ability to monitor, analyze, and respond to the COVID-19 pandemic, ultimately helping to control transmission, mitigate the impact on health systems, and protect vulnerable populations.

8.3.4 Disease Outbreak, Prediction, and Prevention

The team swiftly presented near-accurate answers and built predictive models on disease spread. The model worked on various parameters like reported infected cases, geographic location, socio-economic indicators, hygiene condition, primary health care availability, appropriate drugs, symptoms, and travel data. This was done to identify various critical factors that influence decision-making, such as the area or region to which the citizen belongs, which individuals have more chance of suffering from COVID-19, likely locations of COVID-19 spread, and what steps should be taken to control the spread.

Tracking reported COVID-19 cases provides essential data for monitoring transmission trends, identifying hotspots, and assessing the outbreak's severity. Analyzing case data over time can help predict future trends and guide resource allocation and intervention strategies. Geographic location plays a crucial role in understanding the spread of COVID-19 and identifying high-risk areas. Geographic information systems (GISs) can be used to map the distribution of cases, visualize spatial patterns, and identify clusters or outbreaks in specific regions. This information can inform targeted interventions, such as localized lockdowns, increased testing, and healthcare resource allocation. Understanding the symptoms and clinical manifestations of COVID-19 is essential for early detection, case management, and public health surveillance. Public education campaigns, healthcare provider training, and symptom monitoring tools can help raise awareness, promote early testing, and facilitate timely intervention for suspected cases.

Socio-economic indicators, such as income level, employment status, housing conditions, and access to healthcare, can influence vulnerability to COVID-19 and the ability to adhere to preventive measures. Vulnerable populations, such as low-income communities or migrant workers, may face increased risks due to overcrowded living conditions, lack of access to healthcare, and limited resources for prevention and treatment. Targeted outreach, support programs, and public health campaigns can address these disparities and reduce transmission within vulnerable communities.

Hygiene practices, including hand hygiene, respiratory etiquette, and sanitation, are critical for preventing the spread of COVID-19. Promoting good hygiene through public education campaigns, access to handwashing facilities, and distributing hygiene supplies can help reduce transmission and protect communities from infection. Further, monitoring travel patterns and restrictions can help identify potential transmission sources and prevent the spread of COVID-19 across regions. Travel advisories, border controls, quarantine measures, and testing requirements can be implemented based on travel data and epidemiological risk assessments to limit the introduction and spread of the virus.

By integrating these parameters and leveraging data-driven approaches, public health authorities can enhance their capacity for disease outbreak prediction, prevention, and control, ultimately reducing transmission, mitigating the impact of the pandemic, and protecting public health. The model also looked at predicting trends by analyzing the current situation and past treated data to identify the root cause of issues. This helped to determine how to address performance issues in the

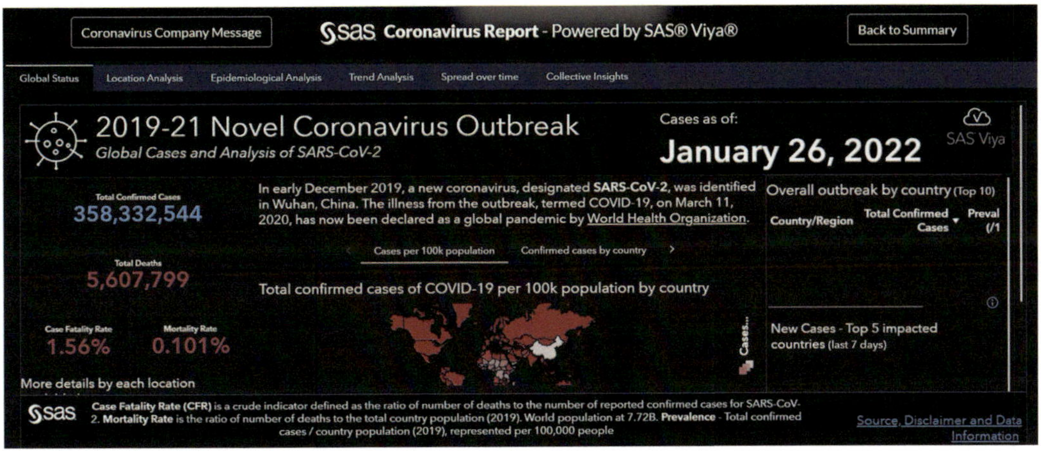

FIGURE 8.2 Global COVID-19 dashboard.

best possible manner. It could help the department to rapidly quarantine and put measures in place to reduce and/or stop the spread of the disease.

8.4 THE COVID-19 DASHBOARD AND ITS IMPACT

This "data for good" initiative has helped various state governments build resilient processes to manage the pandemic using several advanced technologies like data management, statistical analysis, and artificial intelligence (AI). Through collaboration, governments could deploy early warning systems, maintain and display public dashboards at critical moments, and monitor predictive analytics systems for informed decision-making.

The global COVID dashboard developed by Statistical Analysis System (SAS) is a comprehensive data visualization tool designed to provide real-time insights into the COVID-19 pandemic. Leveraging SAS's advanced analytics capabilities, the dashboard aggregates and analyzes COVID-19 data from multiple sources, including government agencies, public health organizations, and research institutions worldwide. The dashboard can be accessed using the following link: Global COVID dashboard (https://tbub.sas.com/COVID19/); a snapshot of it is in Figure 8.2, which provides insights covering the global status of COVID spread. The global COVID dashboard developed by SAS includes location analysis, epidemiological analysis, COVID-19 trends, and collective insights. These serve as valuable tools for monitoring the pandemic, facilitating data-driven decision-making, and supporting public health efforts to combat COVID-19 globally.

NOTE

1 Accessible at https://statedashboard.odisha.gov.in/

BIBLIOGRAPHY

Dong, Ensheng, Hongru Du, and Lauren Gardner. "An interactive web-based dashboard to track COVID-19 in real time." *The Lancet Infectious Diseases* 20, no. 5 (2020): 533–534.

Ghosh, Palash, Rik Ghosh, and Bibhas Chakraborty. "COVID-19 in India: Statewise analysis and prediction." *JMIR Public Health and Surveillance* 6, no. 3 (2020): e20341.

Pathania, Vishali, Sritha Zith Dey Babu, Shahanawaj Ahamad, P. Thilakavathy, Ankur Gupta, Malik Bader Alazzam, and Digvijay Pandey. "A database application of monitoring COVID-19 in India." In *Artificial Intelligence on Medical Data: Proceedings of International Symposium, ISCMM 2021*, pp. 267–274. Singapore: Springer Nature, 2022.

9 Child Health Cluster Analytics in Maharashtra – Non–Key Indicators Tell Key Stories

Raunak Saraf and Anand Kulkarni

ABSTRACT

The Department of Women and Child Development, Government of Maharashtra, focuses on holistically promoting the survival, protection, development, welfare, and participation of women and children in society and the economy. Towards this goal, the Chief Minister (CM) Dashboard, Maharashtra, is created to give a complete bird's eye view to the honourable chief minister of Maharashtra and state officials working in various departments of the state government regarding the health and disability indicators of women and children living in the state of Maharashtra. The dashboard helped in monitoring the key performance indicators defined for each department and its respective initiatives, and thereafter categorised into situational and department reports. Daily and monthly status reports and comprehensive graphic reports are generated for further insight. Insights generated from the dashboard helped the government to plan better and more targeted schemes and policies that combat these specific disabilities in specific districts.

9.1 INTRODUCTION

Maharashtra is the second most populous state in India, with around 11.2 crore citizens per the 2011 Census. Some of its key demographics are that women constitute about 48% of the state population, at about 5.41 crores. Children (0–6 years) constitute around 12% of the state population, at around 1.33 crores. The sex ratio is at 894 for the age group 0–6 years for children. The infant mortality rate is about 19 deaths per 1000 live births.

The main objective of the Department of Women and Child Development, Government of Maharashtra ("Department"), is to promote the survival, protection, development, welfare, and holistic participation of women and children in society and the economy. However, the first criterion for this purpose is to have healthy and mentally sound women and children in the state.

To secure this end, the Department undertakes policymaking, formulation of programs/schemes, implementation of these development programs, and coordination of efforts of both governmental and non-governmental organisations working in the field of women's and child development.

9.2 CURRENT SCENARIO

In this context, the first step to good policymaking is accurately measuring integrated child development based on the available information. Over time, however, given the complexity of the overall ecosystem, keeping track of the progress of existing key performance indicators (KPIs) and comparing them with existing progress of KPIs had become a challenge.

Until now, the general practice has been to visit locations manually and collect data. Therefore, there was often inaccuracy in the accumulated data points. For example, if a child receives benefits

DOI: 10.1201/9781003487531-12

from its visual challenge and health, and the data regarding their nutrition level is unavailable, then the overall growth and child development picture becomes incomplete and unclear. Since the Department could not have accurate and complete data, it was harder to formulate effective schemes and adequately implement them for us to see real change. However, before we arrive at all of this, we need to understand how these discrepancies crept into existing systems.

Earlier, information was collected manually by enumerators. These enumerators would personally make visits to households and collect the data. Therefore, usual human errors involving manual data collection occurred. This, coupled with other issues like moral hazard problems, made the data quite unreliable and, in turn, less useful.

Further, the indicators we had been developing were highlighted based on this noisy data, making our findings somewhat inaccurate. Separately, it must also be noted that the traditional method was very time consuming, and delays often caused significant data-gathering gaps. This was particularly problematic in this scenario, as the data needed was time sensitive, as it was of growing children. These delays adversely affected our policy, making it ineffective in the long term.

Therefore, the major problem was that data collected through traditional methods was unfit, as it was error ridden, inaccurate, and delayed.

9.3 NEW APPROACH

After analysing the situation, the government of Maharashtra devised a smart solution to combat this problem. It was proposed to create a state-of-the-art Chief Minister Dashboard, Maharashtra (CMDM). This dashboard would give a complete bird's eye view to the honourable chief minister of Maharashtra and state officials working in various state government departments regarding the health and disability indicators of women and children living in the state of Maharashtra.

Our aim was for the CMDM to present accurate data and track the overall progress of various departments regarding the schemes they run for the holistic development of women and children in Maharashtra.

After the schemes were formulated, all scheme-specific data would be added to the CMDM database and distributed to our officials throughout the various departments in a user-friendly and easy manner. This would enable our officials to make future decisions for the betterment of society and the citizens of Maharashtra.

To better understand the workings of the CMDM, we present an illustration of the child health cluster analysis.

9.4 CHILD HEALTH CLUSTER ANALYSIS

In case of implementing analytics for the child health cluster analysis, the dashboard of the Women and Child Development Department provides a comprehensive view of the analysed data. The first step to achieve this is to integrate data across all regions of Maharashtra.

Data points include disabilities among children, duration of data, and other relevant parameters. These are presented in the form of statistical graphs within the CMDM. This graphical representation enables our officials to view and manage all data to help them make informed decisions, which, in turn, helps them better plan and execute their related schemes. Thus, our state officials can now continuously evolve their scheme formation and implementation process.

Furthermore, the Dashboard categorises key performance indicators (KPIs) into situational and department reports. This means that first, the Department defines KPIs for itself and for all its various schemes and initiatives. The reports are generated based on the Department's performance in designated areas and situations. A status report is generated along with graphical reports to keep tabs on the progress. Our officials can set daily and monthly reports, depending upon the number of initiatives taken by state officials. The dashboards are also quite user friendly and simple to navigate.

FIGURE 9.1 Sample screenshot of the CM Dashboard, Maharashtra.

In this manner, the Department easily determines what unique schemes can be developed and implemented for public welfare and keeps track of actual execution of those schemes. This has greatly helped us keep track of the multiple schemes and also access their effectiveness.

Separately, the Dashboard has many other interesting and helpful features. Its indicators include filtering (i.e., to sift through the search results) and drill-down (to view more granular and detailed results) so that users can view generic and specific data based on their needs. The Dashboard also contains district-wise data based on lead and lag performances, that is, which district is performing the best and which districts are lagging. This means that state officials can track the progress of different interdepartmental officials and troubleshoot areas where results are not in line with their expectations.

Another remarkable feature of the Dashboard is that the data is automatically updated without any assistance from the users. It must be noted that the frequency of the data updating varies within departments depending on their discretion. Moreover, an SMS alerts/automatically generated SMS alerts facility is provided to the higher authorities to inform them whether timely data update frequency is being met. The departments have mostly agreed to receive real-time updates on the schemes' performance and function. These automatically generated SMSs ensure real-time information updates to the designated officer.

To summarise, some of the salient features of the CMDM are:

Search Option: Filters all the data and reports presented by the Dashboard and customises search results to find exactly what the user wants.

Ranking of Districts: Based on the performance of various districts in terms of certain predefined parameters. This helps create healthy competition among districts and track each area's real-time progress.

It is clear now that the data analysis and features offered by the CMDM help state officials every day in identifying triggering factors that risk the health of women and children and thus curating specific solutions for removing disability and malnutrition amongst the young children of Maharashtra. Screenshots of the CMDM are presented in Figure 9.1.

9.4.1 Sample Dashboard of the Women and Child Development and Integrated Child Development Services

The previous figure demonstrates how the child health dashboard is helping the government to identify various locations, durations, and other relevant factors to measure the current progress of a scheme, which, in turn, helps in defining a future approach toward the overall development

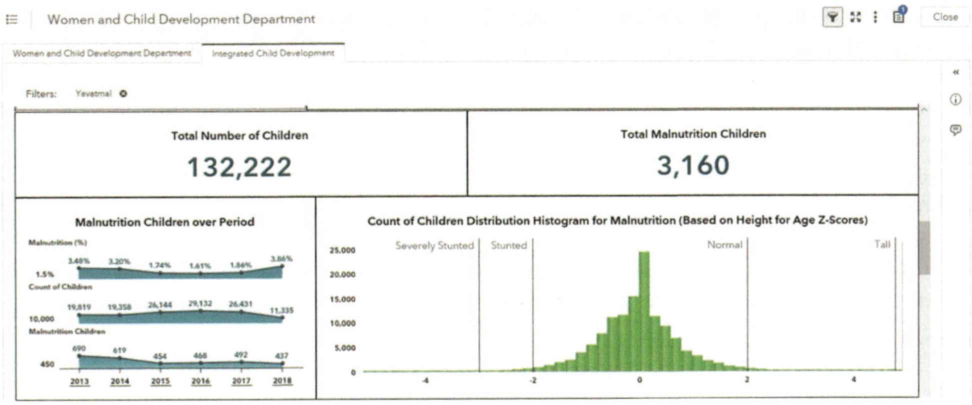

FIGURE 9.2 Dashboard providing a high-level overview of the malnutrition status in Maharashtra.

of women and children in the state. It highlights the top three districts, Satara, Ratnagiri, and Sindhugarh, that have the highest healthy children ratio. The dashboard also provides information on the severity of malnutrition in terms of the distribution concerning age and gender.

Thus, the CMDM helps departments view and analyse malnutrition in children using categories such as location, duration, height, gender, and age group. This analysis not only concentrates on the primary factor of healthy childbirth but also indicates disabilities in kids until the age of 6 years. As an illustration, the figure describes detailed information about child malnutrition in Yavatmal, Maharashtra. Indicators include the number of malnourished children over the years 2013–2018 (Figure 9.2).

9.4.2 Analysis

The overall analysis consists of the following broad steps through the overall duration of the project:

- Government data is collected from various departments and sources
- Data processing involves the ingestion of data into the designated software using rules and necessary aggregation
- Defining the quality of the data for practical and effective insights
- Creation of intuitive dashboards to show stagewise progress
- Meaningful machine-generated insights through intelligent algorithms and statistical methods
- Suggest action plans for the optimisation of the scheme and necessary policy measures

Data type used, collection, and processing: As with most primary data, the Department faced the usual challenges, such as unavailability of relevant data, gathering sufficient data for each indicator, managing and sorting through the overall volume of data collected, and ensuring the upload of authentic data.

The general scenario is that while some data is available with the central and state governments, such data is not always available, often due to a lack of coordination among various departments. In this case, the Department has ensured that this gap does not hinder the Dashboard's processing.

For this analysis, data for the last six years has been used for all age groups across all Maharashtra districts, talukas, and villages. At a disaggregated level, the data has been beneficial for micro-level interventions. The data was shared in flat files consumed by an automated script and uploaded into the Dashboard. The raw data captures a child's height for the age – in ideal scenarios for children receiving adequate nutrition, the child should attain a particular height for a certain age. Looking at this disparate data point across different age groups would make it difficult to compare the data for the entire population. For this purpose, a statistical transformation called the Z-score, a numerical measurement that captures the deviation of values from the overall mean (or average) of the group, is calculated.

This data was thus transformed and processed to calculate the Z-score to calculate the disability index per age-to-height (stunting) ratio. This was compared to mark the Z-score per the categorisation parameters of height, that is, stunted, severely stunted, normal, and tall (worst to best ranking). This parameter was also correlated with the kids' age group and gender to identify which age group and gender were affected the most. (The Z-score measures the extent of under/well nourishment comparing children's height, weight, and age.)

Understanding the overall trend and reasons for disabilities among kids is challenging from such large, segregated datasets, but the Z-score–based analysis helps in identifying the underlying root causes. The Z-score showed that more kids are stunted and severely stunted than taller kids. The analysis also clearly understood the various levels of disabilities, including movement, hearing, speaking, and seeing.

As a final step, the correlation of the scores with the age and gender of the kids with disabilities was done. This helped identify which age group and gender of children are affected by which disability. Once this information is generated, micro-level interventions can be planned effectively. Moreover, this analysis can further correlate with district demographics to identify which factor triggers which disability.

9.4.3 Data Collection and Processing – Challenges and Solutions

This approach can be extended across different departments to derive actionable insights specific to such departments. The various steps in the process are highlighted in the following process diagram. Governments collect a large amount of administrative and operational data, which usually lies in silos across flat files or disconnected databases – collecting this data and segregating it into a single database is the first step in this process.

As these datasets are from disparate sources, quality checks, and data pre-processing form the next important step of this process. The next step consists of generating dashboards on top of the data, which requires defining the key performance indicators in discussion with the departments. Analysis of these dashboards and KPIs helps identify the root causes and areas for improvement. This information can thus be passed in real time to the officers in charge of these activities through SMS or email.

The data is processed and aggregated to be added to the CMDM using the designed rules to ensure its safety and accuracy. At this point, the data quality ensures that the Dashboard can generate effective insights. Once the Dashboard processes the data, it offers insights to aid the optimisation of each scheme rolled out by the government.

9.4.4 Outcome and Impact

What is the overall impact of this dashboard? How has it helped us, the government of Maharashtra, secure the health and safety of women and children of the state? Our analysis of the CMDM reveals that overall, the schemes that were based on the insights generated by our Dashboard have elevated the healthy childbirth rate to 97.73% across the entire state. At the district level, all districts have shown a healthy childbirth ratio of above 95%.

The Dashboard generated some beneficial insights for the state government, such as the statistics of all disabilities occurring in children from 0–6 years (Figure 9.3). Further, movement and speaking were the most identified disabilities. Out of these two, while speaking disability ranked as the highest across most of the districts, movement disability was seen to be the highest in the districts of Aurangabad, Solapur, Ahmednagar, and Thane. These district-specific insights help the government to plan better and more targeted schemes and policies that combat these specific disabilities

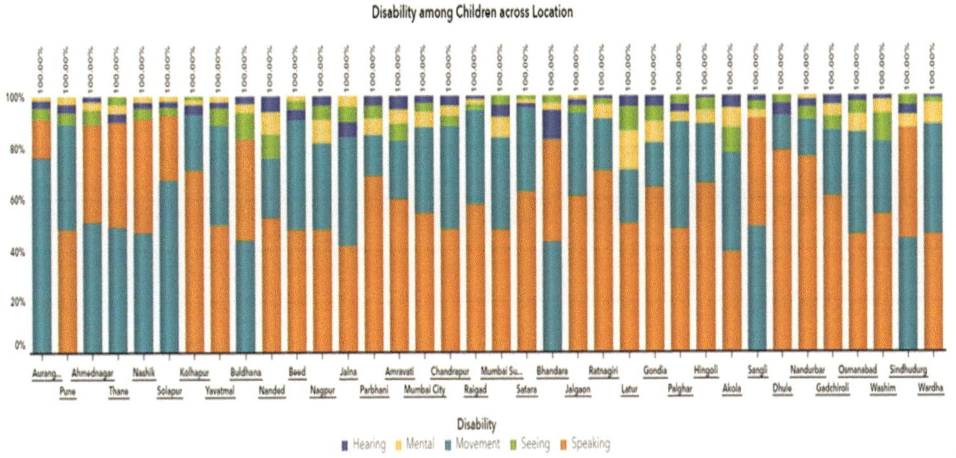

FIGURE 9.3 Dashboard providing details of children's disability across Maharashtra.

in these districts. Therefore, the solution has a lasting impact on improving the health of children in Maharashtra.

First, the KPIs are defined for each Department and its respective initiatives and thereafter categorised into situational and department reports. Daily and monthly status reports and comprehensive graphical reports are generated for further insight.

Moreover, enhanced indicators such as filtering and drill-down are also made available in the dashboard, and they sort these indicators into district-wise lead and lag performances, and so on, so that each department's progress can be tagged. All this data is automatically updated without any assistance from the user. The frequency of the data in the Dashboard and its updating will vary with the departments. Dashboard data gets updated daily. The Dashboard also provides SMS alerts and automatically generated SMS alerts to higher officials over timely data update frequency, as has been agreed by the departments.

BIBLIOGRAPHY

Bennett, Tellen D., Tiffany J. Callahan, James A. Feinstein, Debashis Ghosh, Saquib A. Lakhani, Michael C. Spaeder, Stanley J. Szefler, and Michael G. Kahn. "Data science for child health." *The Journal of Pediatrics* 208 (2019): 12.

Iswari, Lizda, Dhomas Hatta Fudholi, and Silfa Kurnia Aditya. "Dashboarding the maternal and child health profiles for health supporting system." In *IOP Conference Series: Materials Science and Engineering*, vol. 482, no. 1, p. 012013. IOP Publishing, 2019.

Panda, Basant Kumar, Gulshan Kumar, and Ashish Awasthi. "District level inequality in reproductive, maternal, neonatal and child health coverage in India." *BMC Public Health* 20, no. 1 (2020): 58.

Satav, Ashish Rambhau, Kavita Ashish Satav, Abhijeet Bharadwaj, Jayashree Pendharkar, Vibhawari Dani, Suresh Ughade, Dhananjay Raje, and Eric AF Simões. "Effect of home-based childcare on childhood mortality in rural Maharashtra, India: A cluster randomised controlled trial." *BMJ Global Health* 7, no. 7 (2022).

10 Using Data Analytics to Improve Students' Performance

S Anand

ABSTRACT

Surveys are a critical method to collect data about people's behavior patterns to improve human lives. The question analyzed in this chapter is: which behavioral tendencies of students improve their academic performance? Student survey was carried out using a questionnaire of 60 questions divided into four categories: demographic, social, behavioral, and educational preference. This survey is part of the Indian government's National Achievement Survey (NAS) to evaluate the learning outcomes of students and also investigate the underlying factors. Upon analysis, interesting connections were found as discussed in this chapter.

Children who play games perform better at mathematics. Having access to books at home and moderate television watching improves academic performance. Surprisingly, children who do not watch any television perform the worst. Taking private tuition has little to no impact. Finally, all insights were juxtaposed against each other, and the optimal behaviors were derived.

10.1 CURRENT SCENARIO

Anand's neighbor's son, Adhvait, is a precocious 10-year-old. He is into gaming and gadgets. He's glued to the cartoon shows *Chotta Bheem* on TV and *Doraemon* on YouTube. His parents urge him out of the house to play with friends, but he skulks near Anand's door to use the Wi-Fi on his phone.

Adhvait's parents have similar concerns. "Doesn't TV worsen a child's performance?" "Should we have him take up private tuition?" they wonder.

Around this time, Dr Rajaram Sharma, joint director of the Central Institute of Educational Technology (CIET), National Council of Educational Research and Training (NCERT), introduced Anand to a remarkable survey conducted by the NCERT. The National Council of Educational Research and Training is an autonomous organization established by the government of India in 1961. It serves as a premier resource for school education in India, providing academic support and guidance to central and state governments in the formulation and implementation of educational policies and programs. NCERT develops and publishes textbooks, supplementary materials, and educational resources for students and teachers at various stages of school education, from primary to higher secondary levels.

The NCERT has over 20 lakh students across 3rd, 5th, and 8th standards in over 1 lakh government schools across every state. This survey, called the National Achievement Survey (NAS), has two parts. One is competency-based test questions in the general subjects being studied by the students at school. The second is filling out questionnaires about the students, teachers, and schools. It aims to assess student's learning outcomes in various subjects at different stages of school education across the country. NAS provides valuable insights into the education system's effectiveness and helps identify areas that require improvement. The survey covers students from grades 3 to 8

DOI: 10.1201/9781003487531-13

in both government and government-aided schools. It assesses students' proficiency in mathematics, science, language, and social science through standardized tests. The findings of the NAS are used to inform educational policies, curriculum development, and interventions to enhance learning outcomes and the quality of school education in India.

When Dr. Sharma asked if Anand could explore this data, Anand and his team were delighted.

10.2 WHY DO SOME STUDENTS SCORE HIGHER?

We took up the challenge of identifying what student behaviors affected their performance based on two questionnaires:

1. The competency questionnaire, which tested the students in mathematics, reading, science, and social science, and
2. The student questionnaire, which had 60 questions divided into four categories:
    - Demographic information (district, state, gender, age, siblings, etc.)
    - Social information (category, language, parents' income and occupation, below poverty line, etc.)
    - Behavioral information (use of computers, reading books, watching TV, playing games, etc.)
    - Educational preference (subjects chosen, read English, solve math, feel science is difficult, private tuition, etc.)

Which of these behaviors affect the students' marks? By how much? That's the question the team set out to answer.

Before you look at the insights and findings, please understand the nomenclature of the visualization with the following example illustrated in Figure 10.1.

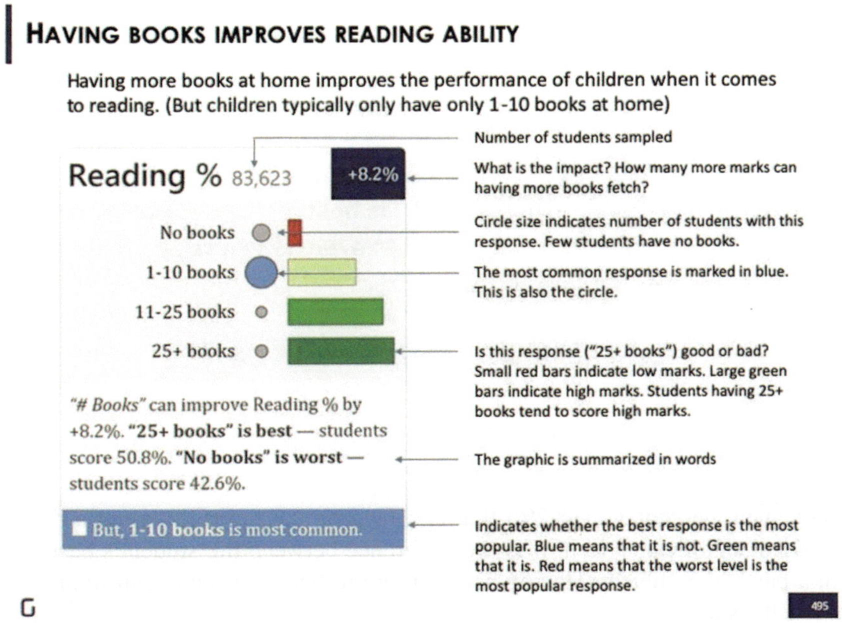

FIGURE 10.1 Explanation of the visualization.

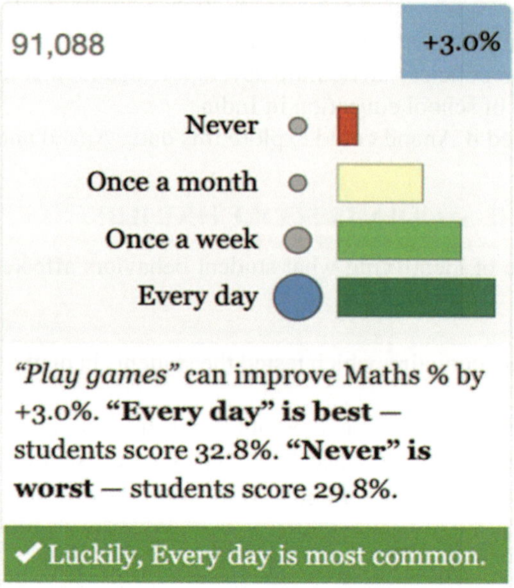

FIGURE 10.2 Visualization showing impact of playing games on mathematics score.

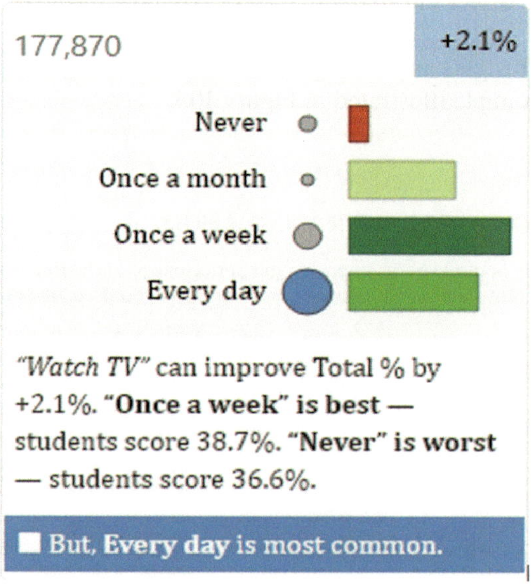

FIGURE 10.3 Visualization showing impact of watching TV on overall score.

The improvement percentage (+8.2%) that is shown on the top right is not a random number. There is statistical significance to it. It is the difference between the student's best and worst performance in a particular subject. Having access to more books at home can improve the average performance of the students.

Let's show another example to explain the improvement percentage. Figure 10.2 shows the impact of playing games on students' mathematics marks.

The improvement percentage is +3.0%. The bars from top to bottom are in increasing order. It means a student who never plays scores 29.8% on average in mathematics (as shown in the figure).

However, students who play games everyday score 32.8%. The difference between them is +3.0%, which denotes the improvement in marks.

10.3 SHOULD WE WATCH TV?

We analyzed the habit of TV-watching behavior.

10.3.1 Impact of Watching TV on Total Marks

It was found that:

- Students who never watched TV scored the least (36.6%)
- Students who watched TV once a month scored 1.3% more (37.9%)
- Students who watched TV once a week scored an extra 0.8% (38.7%) – this is the sweet spot
- Students who watched TV every day scored 0.5% *less* (38.2%) – though this is the most common behavior

To validate these results, the team ran a statistical analysis called the *t-test*,[1] which checks whether the results provide reasons or are merely coincidence. The test showed that there is less than 1 in a billion chance that this conclusion was wrong.

The conclusion was clear. TV watching is beneficial if done in moderation.

10.3.2 Should We All Watch Less TV?

Most children (67%) reported that they watch TV every day. This is suboptimal. So, should they all cut down on their TV watching? Will this help them in every subject? The analysis in this section helps us answer this question.

It certainly holds true for the subjects of science and social science. This conclusion is illustrated in Figures 10.4 and 10.5. Students who watch TV *every day score less,* but students who watch TV once a week score higher.

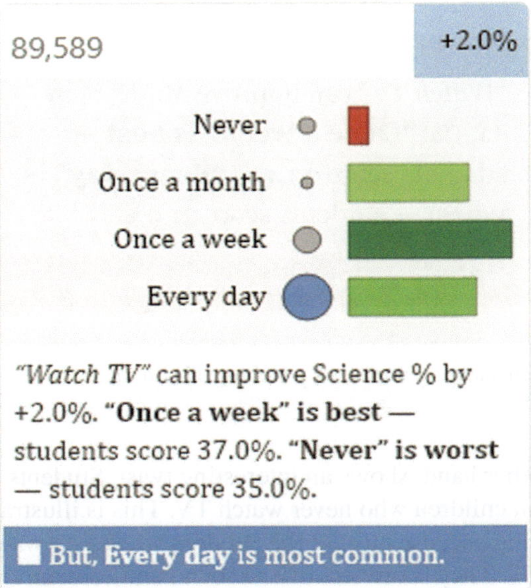

FIGURE 10.4 Visualization showing impact of watching TV on science score.

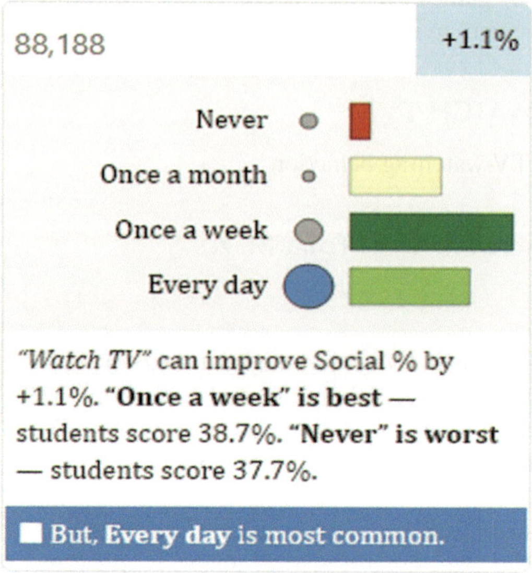

FIGURE 10.5 Visualization showing impact of watching TV on social science score.

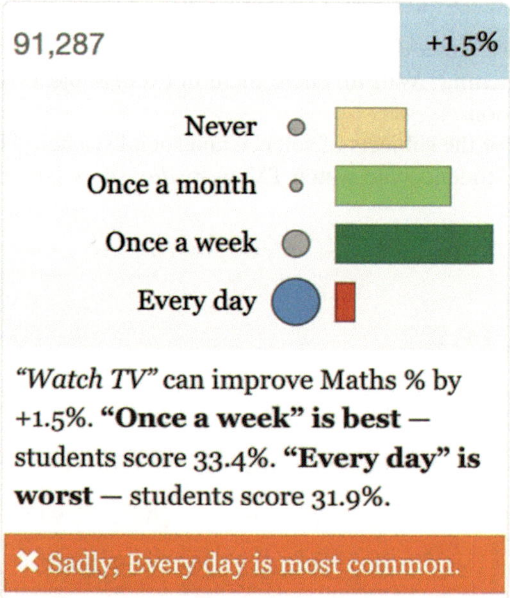

FIGURE 10.6 Visualization showing impact of watching TV on math score.

Mathematics, on the other hand, shows an interesting twist. Students who watch TV daily score the worst, even worse than children who never watch TV. This is illustrated in Figure 10.6.

This discovery is striking because most of the children who never watch TV do not because their households *don't own a TV*. They score less due to social and economic reasons. But children who watch TV daily score *even less* than these underprivileged children.

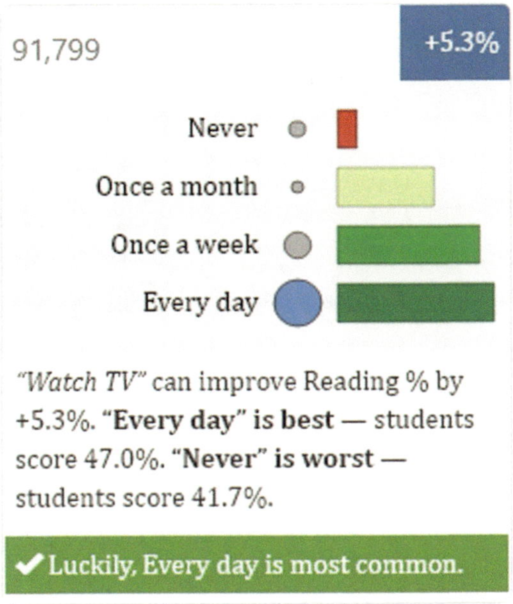

FIGURE 10.7 Visualization showing impact of watching TV on reading score.

Therefore, watching TV every day destroys mathematical ability in students, a result that is again statistically significant. However, there is a particular competency that comes to life with watching TV – reading ability. This is illustrated by Figure 10.7.

The survey found that children who watch TV daily score 5.3% more than children who never watch TV. This is a clear and strong improvement in children's reading ability.

Therefore, the message seems to be twofold:

- Watch more TV to improve your reading ability.
- Watch less TV to improve your mathematical ability.

10.4 SHOULD CHILDREN PLAY LESS OR MORE?

When analyzing playing behavior, we found that children who never play score lower across subjects. But if they play at least a little, their scores are fine. However, how much they play does not significantly improve their scores. This can be found by reflecting on Figure 10.8.

10.4.1 Impact of Playing Games on Total Marks

But, as with watching TV, there is a twist when it comes to playing. Children who play daily do much better in mathematics. They score 1% more than those who play once a week. Those who play once a week, in turn, score 1% more than those who play once a month. Incidentally, those who play once a month, in turn, score 1% more than children who never play.

10.4.2 Math and Readability – Conflicting Conclusions

If we go to Figure 10.2, we conclude that playing has a good impact on improving mathematical ability. But playing more reduces readability skills in children, as depicted in Figure 10.9. Those who play every day score 1% less than those who play once a week, and so on.

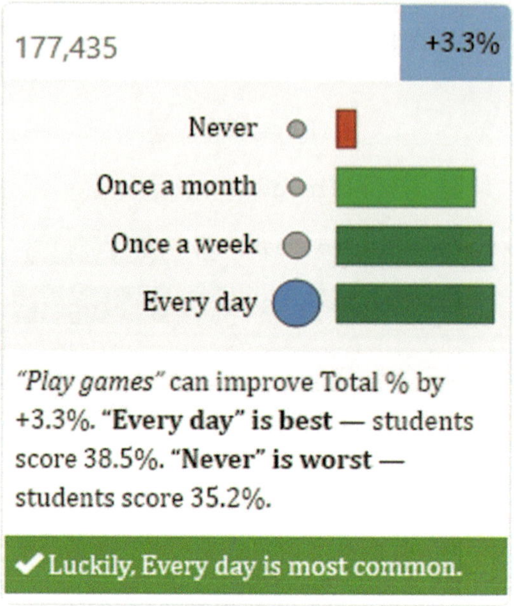

FIGURE 10.8 Visualization showing impact of playing games on total score.

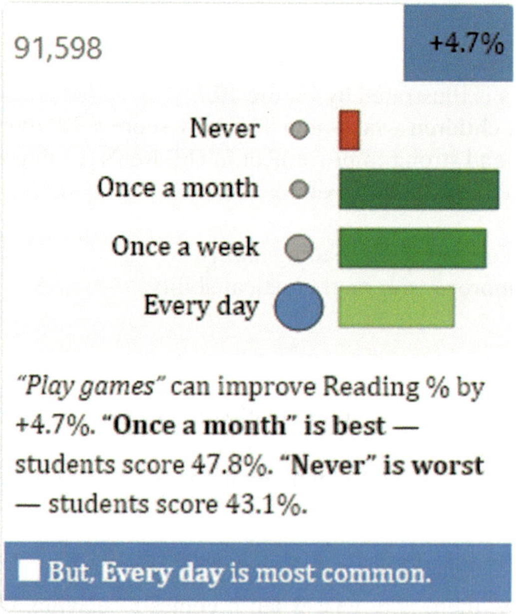

FIGURE 10.9 Visualization showing impact of playing games on the reading score.

10.4.3 Impact of Watching TV and Playing Games on Marks

The message seems to be:

- Play games more to improve your mathematical ability.
- Watch more TV to improve readability.

Factor	Total %	Maths %	Reading %	Science %	Social %
Watch TV	2.1%	1.5%	5.3%	2.0%	1.1%
Play games	3.3%	3.0%	4.7%	4.0%	2.9%

FIGURE 10.10 Comparing effects of watching TV and playing games on student scores.

Let's circle back to Advait and his struggle. With this, Anand and his team were able to advise Advait's parents.

"If your son wishes to be an engineer, turn off the TV and ask him to play more. If he wishes to be a journalist, turn on the TV and guide him as he watches."

10.5 WHAT AFFECTS SCORES THE MOST?

When we compare the overall impact of watching TV and playing games, we found that in almost every area, playing games has a more substantial impact, except readability. Here's a comparative analysis presented in a tabular format (Figure 10.10).

This can be thought of in terms of the relative impact of watching TV and playing games.

For example, the right kind of TV-watching behavior can improve the total by 2.1%, while the right game-playing behavior can improve the total by 3.3%. This gives us a hint about which behaviors are more effective and which recreations we should drive our children towards.

The relative impact of all the factors that significantly impact children's scores can be seen in Figure 10.11.

10.5.1 All Factors Combined – Relative Insight

There are several interesting inferences here.

- The *parents' education and occupation* have the strongest influence.
- On the other hand, *private tuition* seems to be trivial.
- *Using computers* improves mathematics marks significantly. Using calculators, however, does not.
- Having *siblings* does not impact marks much but does have a big impact on reading.

Let's explore these in detail.

10.5.2 How Much Do Parents Matter?

The impact of the father's education can be seen in Figure 10.12.

10.5.3 Father's Education and Total Marks

Ranging from illiterate fathers to those with a degree, the improvement in marks is factored at 11%. In the following, we can see the impact of the father's occupation. See Figure 10.13.

10.5.4 Father's Profession and Total Marks

Children of teachers and professionals again outperform the children of unemployed fathers by almost 11%. Parents are still the single most considerable influence over a child's future success, but

Factor	Total % ▼	Maths %	Reading %	Science %	Social %
Father edu	11.0%	6.6%	18.8%	9.9%	7.9%
Father occupation	10.6%	8.7%	17.4%	10.3%	8.5%
Mother edu	10.5%	4.3%	18.4%	10.3%	7.7%
Mother occupation	9.1%	5.7%	14.9%	7.3%	7.9%
Help in household	5.2%	4.6%	5.7%	5.0%	6.5%
# Books	4.7%	2.5%	8.2%	4.1%	3.8%
Age	4.1%	3.1%	8.0%	3.3%	2.9%
Read a book	3.7%	2.0%	7.1%	3.5%	2.5%
Read magazine	3.4%	1.6%	7.2%	2.3%	2.4%
Computer use	3.4%	6.5%	5.9%	3.2%	4.1%
Play games	3.3%	3.0%	4.7%	4.0%	2.9%
Use dictionary	3.3%	1.3%	6.4%	3.1%	2.2%
Below poverty	3.3%	1.8%	5.7%	2.6%	3.0%
Read other books	3.1%	0.8%	6.4%	2.5%	2.6%
Library use	2.8%	2.9%	5.6%	2.7%	3.3%
Use calculator	2.7%	0.9%	5.0%	2.7%	2.1%
Siblings	2.4%	1.5%	8.3%	2.2%	0.8%
Watch TV	2.1%	1.5%	5.3%	2.0%	1.1%
Private tuition	1.3%	1.3%	2.1%	1.2%	0.3%
Distance	0.8%	2.2%	1.2%	0.9%	0.9%
Gender	0.7%	0.2%	1.9%	0.1%	0.5%

FIGURE 10.11 Relative impact of various factors on student scores.

it is equally interesting to note that certain factors that are stressed a lot by parents turn out to be relatively unimportant. The biggest example of this is private tuition.

10.5.5 Does Private Tuition Really Make a Difference?

With data, the challenge is always to locate non-obvious insights. NCERT conducted surveys to acquire information on students and their family backgrounds. However, NCERT faced a huge roadblock with no predictive and prescriptive results from the collected data.

For example, one of the biggest challenges for NCERT was to prove that tuition culture does not add value but adds stress to their students. However, convincing parents of this fact without any solid data insight was difficult. While analyzing the survey data, we found that the impact of tuition on students' overall marks is almost negligible.

In fact, private tuition improved students' total marks by only 1.3%. Surprisingly, the impact of private tuition on students from below-poverty-line families was merely 0.7%. The subject-wise impacts of private tuition for above-poverty-line and below-poverty-line families are shown in

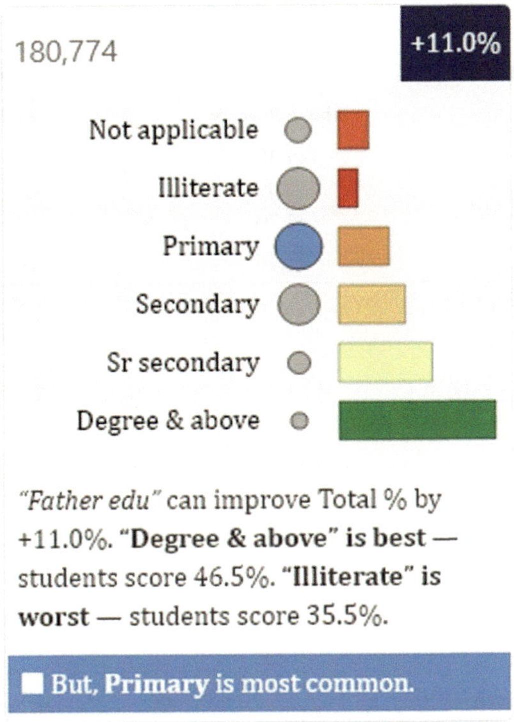

FIGURE 10.12 Visualization showing impact of father's education on the total score.

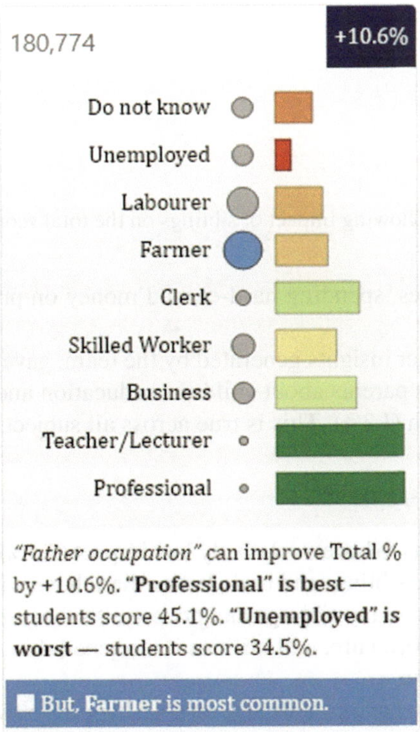

FIGURE 10.13 Visualization showing impact of father's profession on the total score.

TABLE 10.1
Showing the Impact of Private Tuition on Scores across Subjects

Subject	Above Poverty	Below Poverty
Math	+1.3%	+1.3%
Reading	+3.0%	+0.5%
Science	+1.7%	+0.8%
Social	+0.5%	+0.1%

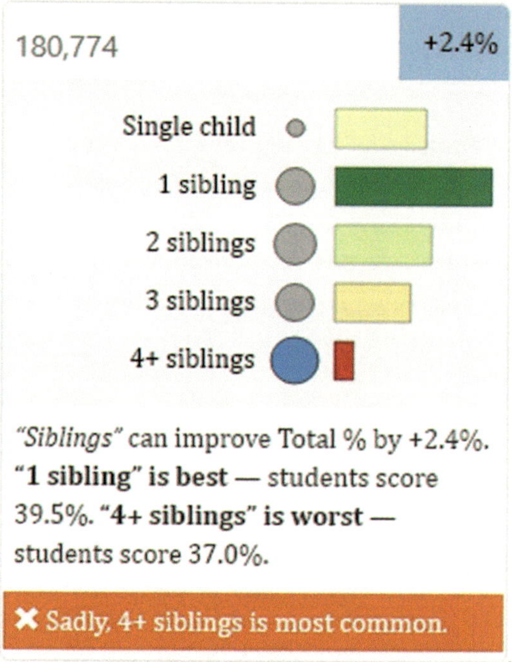

FIGURE 10.14 Visualization showing impact of siblings on the total score.

Table 10.1. For such categories, spending hard-earned money on private tuition hurts more than it helps.

This, along with many other insights generated by the team, gave significant leverage to NCERT when initiating dialogue with parents about children's education and future success. Private tuition does not improve marks much (1.3%). This is true across all subjects.

10.5.6 Does Family Size Matter?

The presence of siblings seems to have a relatively low impact (2.4%) on children's academic performance. Children with a single sibling tend to do better than those with more siblings. This could be due to the correlation between many siblings and poverty. As the number of children in a household increases, the supply of comfort, care, and utilities is stretched thin.

But there is a twist here, too.

When it comes to siblings and marks, a strange reversal happens only for mathematics. The more siblings a child has, the better they seem to perform in mathematics. A detailed look is given in Figure 10.15.

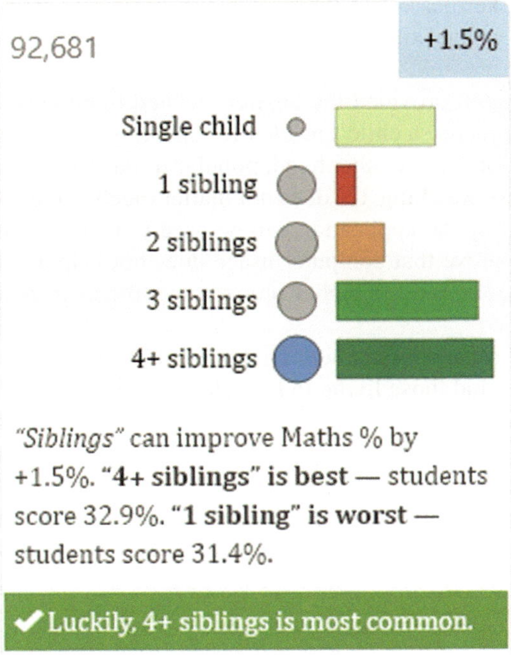

FIGURE 10.15 Visualization showing impact of siblings on the math score.

10.5.7 Impact of Siblings on Mathematics Marks

Interestingly, this is true only of mathematics. Unfortunately, the data only indicates the trend. There isn't yet an explanation for this disruptive behavior. Perhaps children with more siblings learn how to divide better?

10.5.8 State-Wise Analysis: Are Parents the Key Determining Factor in Every State?

It was broadly noted that the parents' occupation and education are the dominant factors in every state. However, it has a few quirks.

For example, the father's education and occupation are bigger drivers within the country overall. However, in a patriarchal state like Punjab, the mother is more important than the father. The mother's occupation has a much higher impact (16.8%) than the father's occupation (6.5%)

On the other hand, in a matriarchal state like West Bengal, the father's influence is higher. The father's occupation has a 20.6% impact – much higher than the mother's (8.3%).

This may be due to the mother's primary responsibility for children's education in Punjab, while in West Bengal, it may be the father. Kerala is another state where the impact of the father's occupation is much higher than the mother's, while in Haryana and Andhra Pradesh, the mother's occupation matters more than the father's.

10.6 CONCLUSION

An important problem today is the availability of data. The National Achievement Survey solves this problem by running a large-scale survey. But how do they interpret and analyze this big data?

We solve challenges like these. The interactive application we've deployed aims to solve this challenge. Policymakers wanted the answer to one specific question: What behavior *affects students' marks the most?*

Our interactive application provided the answers related to parents' education and occupation. Among the factors that influence a child's academic capability is reading more books. This, by far, is the most significant factor. On the other hand, popular trends like private tuition make little or no difference. Watching or not watching TV does not matter much, either.

The insights generated by the application can be used to help decide between alternatives. For example, there is data to prove that computer usage does not help as much as reading books. So, given the lower cost of books and the higher return, investments in books could be a much better initiative than in computers.

This can be sharpened further by segmenting. For example, does this hold true for children from below-poverty households and those living in households above poverty? It turns out that for those below poverty, computer usage is *more important* than reading books. So, investment in computers for schools serving those below poverty levels is a more critical initiative so that these children get access to computers from an early age.

Exposing policymakers to a tool that helps answer these questions through analyzing big data leads to better decision-making backed by not just experience but also solid data. Children are the future of our country. It's critical that we analyze the data we have to influence their academic well-being in the right direction.

NOTE

1 A *t*-test is a statistical method to determine if two sets of data are significantly different. It compares the two sets of data and their means. The *t*-test assumes the data follows a normal distribution and that the variances of the two groups of data are equal. The test uses a null hypothesis, which states no difference between the groups, and an alternative hypothesis, which states otherwise. The final value resulting from a *t*-test is called the *p*-value. It tells us how likely it is that the difference between the two groups could have happened by accident. If the *p*-value turns out to be less than 0.05, it proves the alternative hypothesis.

BIBLIOGRAPHY

Johnson, Doug, and Andres Parrado. "Assessing the assessments: Taking stock of learning outcomes data in India." *International Journal of Educational Development* 84 (2021): 102409.

Kyriakides, Leonidas, Christiana Christoforou, and Charalambos Y. Charalambous. "What matters for student learning outcomes: A meta-analysis of studies exploring factors of effective teaching." *Teaching and Teacher Education* 36 (2013): 143–152.

Trigwell, Keith, and Michael Prosser. "Improving the quality of student learning: The influence of learning context and student approaches to learning on learning outcomes." *Higher Education* 22, no. 3 (1991): 251–266.

11 Determinants of Placements in the Indian Skilling Ecosystem

Dipyaman Sanyal

ABSTRACT

The National Skill Development Corporation (NSDC) fulfilled its primary objective of supporting the skilling system and upgrading skills to international standards with the help of a data analytics pyramid. A large amount of data was fed into the data analytics pyramid from the bottom level, that is, the status of skills, and then moving upwards to diagnose, predict, and prescribe upskilling principles.

The dashboard offers visual analytics (which can provide key performance indicators to decision makers) and at the same time inputs for the models that can be utilised for predictive and potentially prescriptive analytics. All data is completely anonymised and then processed to generate valuable data insights such as placement rates across age groups, states, sectors, and industries. Key insights were generated around formal and informal placement rates, course length, and technical education of candidates.

11.1 INTRODUCTION

> National Skill Development Corporation (NSDC) is a not-for-profit public limited company incorporated on July 31, 2008, set up by the Ministry of Finance as Public Private Partnership (PPP) model. The government of India through the Ministry of Skill Development & Entrepreneurship (MSDE) holds 49% of the share capital of NSDC. In comparison, the private sector has the balance 51% of the share capital.[1]

With over 462 training partners and 11,000+ training centres, working across 2100 job roles and 38 sectors, and having trained over 1.12 crore (11.2 million) participants (with an approximately 50% placement rate),[2] NSDC has provided an unprecedented boost to the Indian skilling ecosystem.

11.2 CURRENT SCENARIO

In keeping with the primary mandate of NSDC, which is to upgrade skills to international standards in different sectors by enhancing and supporting private sector initiatives while trying to narrow the gap between the demand and supply of skills, it can be argued that one of the key focus areas internally for the organisation would be to analyse the current scenario of placement rates. This includes observing the success rates of placements in terms of factors such as age, gender, education level, sector, location, and the interaction effects of each of these factors and many others.

These can determine the significance of the previous factors in increasing placement success and help NSDC decide where they can have the most significant potential social and financial impact on their trainees. The effects of different determinants on the placement rates would help provide essential knowledge on how to take vital decisions in terms of skilling, which would help place a specific individual on the path to success. Courses and centres can be tailored for an individual based on the information provided during enrolment. It should be noted that NSDC, through its funding and as the custodian of specific government schemes, supports the skilling ecosystem via the open markets.

DOI: 10.1201/9781003487531-14

The organisation provides loans to training partners with specific financial and social goals. Still, the final absorption of the courses provided by the partners is driven by the demand and supply dynamics of the market. Thus, it may be the case that analysis suggests a significant shortfall in a specific sector, which is bound to lead to potential employment opportunities – and NSDC funds training partners in that sector in the relevant geographies. However, from the market's supply side, there need to be enough students willing to take on that course who can then be a part of the labour supply in that sector. This matching is essential to NSDC's plans and its mandate to use skilling for the financial uplift of the many.

11.3 BASIC AND ADVANCED ANALYTICS FOR SKILLING

Data analytics can help in achieving each of these goals. The knowledge and experience of policy makers, user data, and analytics provide various insights into the direct and indirect linkages between skilling and placement rates. While the focus of most analytics across organisations is on descriptives and diagnostics (Figure 11.1), NSDC is an early adopter of advanced analytics in some areas of its data since it was evident from global research on vocational training that with the help of data visualisation and machine learning models, a deep dive can be made to enhance the success rates of placements based on myriad factors which can be ascertained in the process of implementation of such analytics.

As any data scientist knows, data cleaning and processing take much longer than modelling the data. Similarly, understanding the business, understanding the internal data sources and external potential data connections, and then building up a data set to work on often takes longer than the modelling phase. The most crucial step before a statistical or machine learning model can even be implemented is data exploration.

Exploratory data analysis (EDA) overlaps with descriptive and diagnostic analytics. It can often be considered the final step of an analytics activity before inferences are drawn from the data. This phase is highly time consuming since analysts must look at different aspects of the data, slice and dice the data based on business understanding, and then come up with a dashboard for visual analytics (which can provide key performance indicators to decision-makers) and at the same time inputs for the models that can be utilised towards predictive and potentially prescriptive analytics.

The first challenge for any such project is data security and privacy. NSDC is the repository of data for multiple skilling programs. Still, this data is susceptible and must be masked entirely and anonymised before a medium-sized random subset of the data can be provided. Two groups of students were provided 100,000 data points of individual participants randomly selected from the

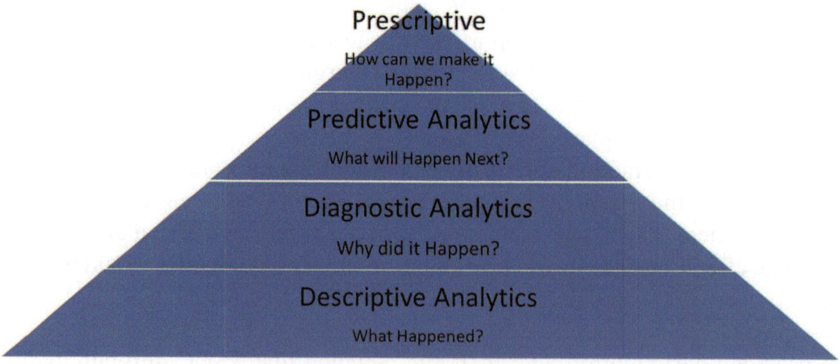

FIGURE 11.1 The analytics pyramid. While most organisations focus only on the bottom tiers of the pyramid, NSDC reaches the top while not ignoring the foundations.

historical skilling database – one from the fee-based (FB) NSDC programs and the other from the Pradhan Mantri Kaushal Vikas Yojana (PMKVY). The intent was twofold:

1. The students would create an interactive dashboard that internal stakeholders could use to analyse the data and create different cuts of the data, which would automatically be visualised using relevant graphs, charts, and tables.
2. The students would use machine learning algorithms to estimate effects of different determinants on placement rates. This would also be incorporated in the dashboard to allow users to easily visualise the outcomes of the models and use it for their own inferences.

To achieve this goal, the entire dashboard would be created in an open-source R-Shiny application, which allows for easy integration with internal websites and can also scaled up to be used on the millions of data points in the SDMS (data management system), while only a small subset would be used by the students to create it. And, as an effect of these steps, the open-source placements dashboard could be used by internal stakeholders and for decision makers to slice and dice the data, understand different KPIs, and understand performance in terms of placements by sector, state, education level, gender, and multiple other factors.

One set of students focused on creating a dynamic dashboard using R-Shiny for the top management team to help provide them with a quick look at key factors for successful training efforts undertaken by NSDC. The dynamic dashboard can help them take important decisions such as which centres/courses/packages are more helpful for higher placement success. It also provides insight into essential parameters for people enrolled and their performance so that courses/counselling can be designed in a specific manner accordingly. An employment prediction model was also created to help ensure the person takes the right course to ensure maximum placement success for the candidate.

The dashboard consists of sections (as tabs) provided to the user to perform data analysis operations and derive insights:

1. *Data Preparation*: where users can upload their dataset for cleaning and preprocessing. There is an option to perform missing value treatment based on the type of column selected. A "View" button displays the distribution graph for each feature (see Figure 11.2

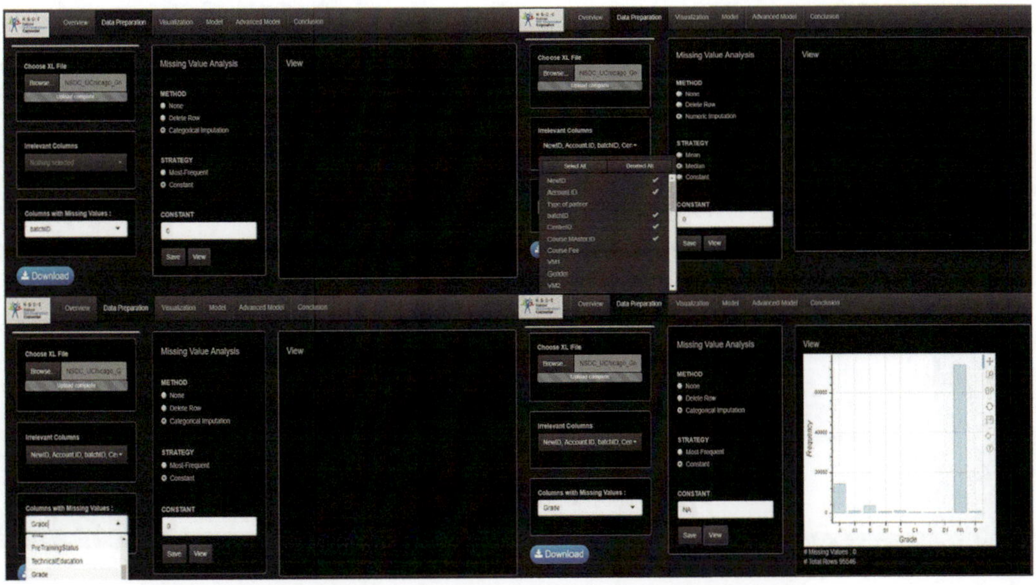

FIGURE 11.2 Showcasing the four data cleaning process tabs – from ingestion and input to graphing.

for the detailed steps which work through data import, dropping of irrelevant columns, missing value input if needed, and basic graphing for a column).

2. *Visualisation*: Visual charts that show important statistics. Dynamic interactive charts for the users to compare and derive helpful insight. This can be seen for a small subset of the NSDC data shared with the students in Figure 11.3. Please note that since the data was a small cut of the entire data, it does not represent the overall values – and since the data was anonymised, most variable names are masked.
3. *Model*: A decision tree algorithm is run on data to predict the probability of placement success. This tab also contains a feature importance graph, the model receiver operating characteristic (ROC) curve, and validation results. These help in providing the user with essential information regarding model success
4. *Advanced Model*: User can choose random forest or an XGBoost algorithm to run the predictions. The ROC curve plots, accuracy, and feature importance tabs are provided to compare and choose the best model. Figure 11.4 showcases both the Modeling and Advanced Modeling tabs. The outcome of these tabs is not only the model itself but a visualisation of the tree (in the case of the decision tree model) and feature importance (for all models) to help understand the underlying dynamics of the multiple models and see the drivers and factors and their quantitative strength in predicting placements.

Random forest is an ensemble learning method that constructs multiple decision trees during training, where each tree is built using a random subset of the training data and a random subset of the features. During prediction, each tree independently predicts the output, and the final prediction is determined by aggregating the predictions of all trees (e.g., by taking the majority vote for classification tasks or the mean prediction for regression tasks). Random forest is known for its robustness to overfitting, versatility across different datasets, and ability to handle high-dimensional data efficiently. It's widely used in practice due to its simplicity, scalability, and competitive performance across a variety of machine learning tasks.

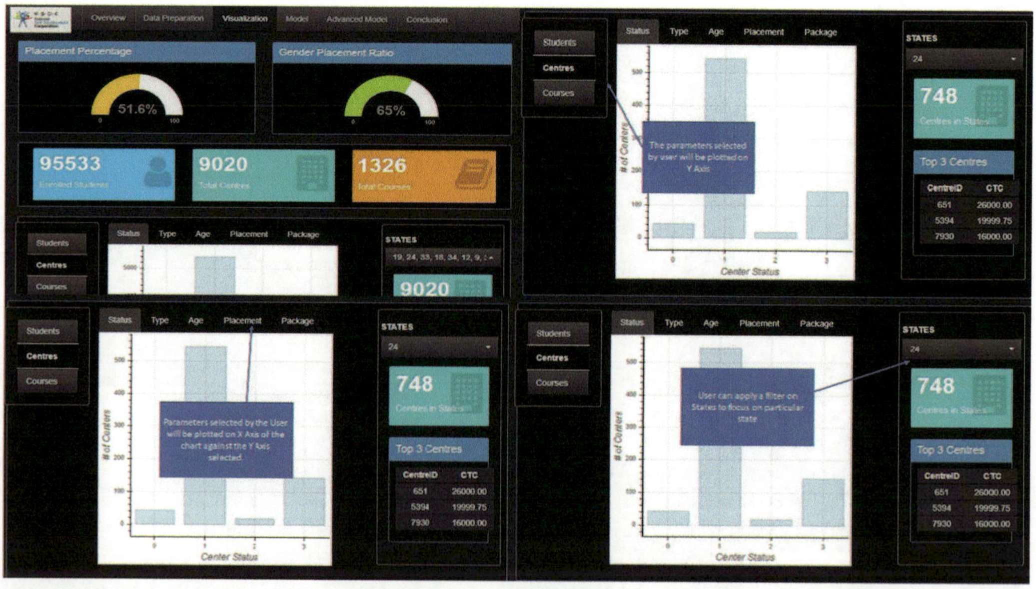

FIGURE 11.3 Showcasing the four visualisation process tabs – from dashboard to data/axis selection and filtering.

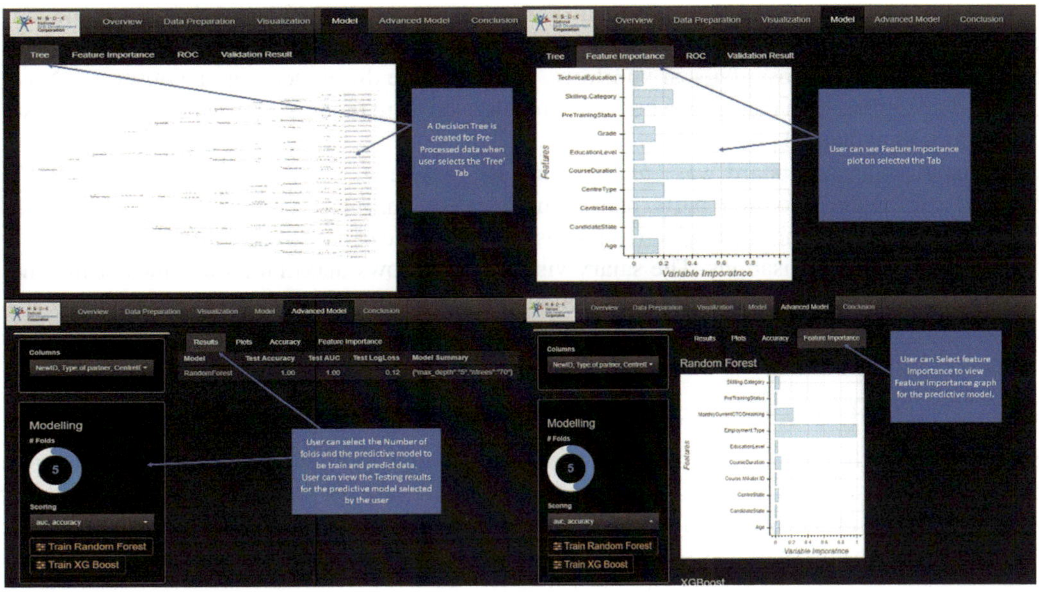

FIGURE 11.4 The modelling tabs – from model and parameter selection, tree diagram, and feature importance.

XGBoost is a gradient boosting algorithm renowned for its speed, scalability, and performance. It builds a sequence of weak learners (typically decision trees) iteratively, with each new learner attempting to correct the errors made by the previous ones. XGBoost optimises a differentiable loss function at each step, incorporating regularisation techniques such as shrinkage (learning rate) and maximum depth of trees to control overfitting. It supports parallel processing and efficiently handles missing data, making it suitable for large-scale datasets. XGBoost often achieves state-of-the-art performance in various machine learning competitions and is widely adopted in industry due to its predictive accuracy and versatility.

The receiver operating characteristic curve is a graphical representation commonly used to assess the performance of binary classification models. It plots the true positive rate (sensitivity) against the false positive rate (1-specificity) for different threshold values. A diagonal line represents a random classifier's performance, while a good classifier's ROC curve typically lies above this line, indicating higher sensitivity and specificity. The area under the ROC curve (AUC) quantifies the model's discriminative ability; an AUC closer to 1 suggests better performance. ROC curves are valuable tools for comparing and selecting models, especially when the class distribution is imbalanced or the cost of misclassification varies.

Some of the insights garnered from data analysis and the machine learning model outputs were on certain factors such as age – the maximum number of students are between the age group of 18–30, and there is maximum enrolment of courses within this age group. The dashboard provides graphs of top courses per state as well as the topmost earning skill set.

Another group of students worked on understanding what the exact determinants of placements were. Exploratory data analysis was done extensively, showing different features and their behaviour in terms of placements. One set of models was built to predict the probability of placement success. For this, different classification algorithms, such as logistic regression, decision tree, K-nearest neighbour, Gaussian naive Bayes, and random forest (for explanation, refer to the technical appendix at the end), were run and their accuracy scores were compared to select the best model. Another set of models was created to observe the monthly income analysis. From the exploratory data analysis, it was seen that the median salary across different age groups was pretty much in the same range, which indicates that outliers are present. Median salary is higher in the older age groups, which is reasonable. Salaries across states are uniform, and no severe differences can be observed. No disparity of salary based on gender can be seen. Models used were random forest and gradient boosting algorithms, as explained previously.

A visualisation dashboard was created using R-Shiny, a standard tool for data analysis. The dashboard is divided into three parts: placement visualisation, salary visualisation, and the models. For the visualisations, there are selection panels that the user can use to enter different variables and observe different visualisations. The salary visualisation shows maximum outcomes of income in certain features. The model tab provides options for the user to explore and run predictions using different algorithms. Model features are also displayed to make comparisons.

11.4 INSIGHTS FROM THE DATASET

Since the size of the data shared with the participants was quite limited, the insights from the project may not hold for the entire NSDC data. However, certain patterns in the data were quite evident. And, while most of the key insights cannot be made public due to confidentiality clauses, the following can be broadly shared as some of the patterns seen in the limited data which was shared:

1. The formal vs. informal sector placement rates for the same industry might be significantly different. It has been found that in certain cases, due to the target of NSDC skilling programs, informal sector placements in some sectors tend to be higher than more advanced formal sectors.
2. State-level differences in placement rates for individual sectors are often quite similar. While we know that state-level development of certain sectors is different, this can be a sign of inter-state mobility.
3. Course duration often tends to be one of the most important determinants or predictors of placements, but what is often expected, that technical skills are a large determinant of placements, does not hold true in this data. Technical vs. non-technical skills have quite similar placement rates, and, other things remaining constant, technical skills do not predict a better placement probability.

11.5 THE DATA INITIATIVE

While my role as a consultant with the market analytics team at NSDC focused on collecting relevant internal and external data and understanding the business to provide valuable insights to it, the realisation that some support on the EDA phase would be extremely helpful was obvious to us after a few months into it.

The team worked closely with the NSDC Market Analytics team, a small but extremely efficient group. At the same time, there was a joint effort with the University of Chicago–Jigsaw Academy and Industry Partner, IBM.

It was clear that bringing the data scientists and the NSDC team together would effectively be a win-win situation for both NSDC and the data team. The synergies were obvious – the data team would get to work on real-world problems as a part of this initiative. In contrast, at NSDC, if we could guide the work properly, we would get some insights and exploratory data analysis by a well-trained group of students. Additionally, the team were trained on open-source technology like Python and R, which could help NSDC replace some of the paid tools used with more flexible and free resources. The leadership team at NSDC and the Market Analytics team realised the value of the potential partnership, and within a very short period, a non-commercial MoU was signed to kick off the project.

11.6 IMPACT OF THE INITIATIVE

The potential of analytics is ever expanding. There is room to inculcate more techniques to unleash the full power of analytics that further help understand and enhance placements' success while monitoring continuous skilling. For instance, incorporating deep learning into the predictive domain would further improve the predictions and provide better results in estimating the effects of placement determinants.

As we delve more into the different phases of exploratory data analysis, we are exposed to specific insights that may cause a shift in the way we perceive and understand the data in this domain. EDA is highly significant when it comes to using machine learning models. The sort of intuition that one may have on specific factors can be further grounded with the help of such techniques. EDA helps uncover the significance of certain determinants that play a key role in prediction.

This knowledge, combined with the vast amount of machine learning, advanced deep learning, and cognitive learning algorithms that are available to us, is a step towards further understanding what enables success and progress in all sectors. With the democratisation of machine learning and the rise of explainable AI, the path that leads to understanding and using analytics only gets better from here on out.

NOTES

1 NSDC website. Accessed on 30 April 2020. https://nsdcindia.org/about-us
2 NSDC website. Accessed on 30 April 2020. https://nsdcindia.org/about-us (data as of July 31, 2019 reports).

BIBLIOGRAPHY

Chenoy, Dilip. "Skill development in India a transformation in the making." In *India infrastructure report 2012*, pp. 199–207. Routledge India, 2016.

Gupta, Deepa, and Sugandha Agarwal. "Skill development initiative-literature review." *Inspira-Journal of Modern Management and Entrepreneurship (JMME)* 8, no. 2 (2018): 319–322.

Gupta, Rajesh, and Oshin Dharap. "How is India skilling its youth? A comprehensive study." *Journal of Vocational Education & Training* 76, no. 4 (2024): 818–844.

Section III

Agriculture, Climate and Environment

12 Decision Support System for Sustainable Growth in the Agriculture Sector

Chandrakiran Mallarapu, Nikhilesh Kumar, Aswin Kumar, and Saptarshi Mondal

ABSTRACT

An integrated digital solution, the Andhra Pradesh Agriculture Information and Management System (APAIMS), was developed by the government of Andhra Pradesh to analyze various agricultural data sets and produce meaningful advisories. The primary stakeholder is the farmer; all advisory and support services are created and offered with the ultimate benefit of the farmer in mind.

The integrated platform captures data such as soil health card data, localized weather data from automatic weather stations (AWSs), and soil moisture data from sensors, and further, it superimposes insights derived from satellite data, information on water availability, geospatial data, crop phenology models, and so on. The APAIMS digital platform capabilities include assimilation of complex data backed by analytics and a knowledge engine to produce holistic "farm-specific" advisories.

12.1 INTRODUCTION

India, with its rich agricultural heritage and diverse agro-climatic zones, stands as one of the world's leading agricultural nations. Agriculture has been the backbone of India's economy for centuries, contributing significantly to employment, food security, and rural livelihoods.

Per the Ministry of Statistics and Programme Implementation (2018–2019), the share of agriculture and allied sectors to the gross value added (GVA) of the country at current prices stands at 15.87%, while the services sector and industry sector contribute to 54.40% and 29.73% of India's total GVA. The contribution of agriculture and allied sectors to the GVA has been steadily decreasing.

The government of India announced in 2015–16 its goal of doubling farmers' income by year 2022, which will require an annual growth rate of 10.4%. However, the growth of the agriculture sector in the last few years has been hovering around slightly more than 2%.

The challenge faced by the agriculture sector in India is the stagnation of productivity levels compared to global standards, owing to small land holdings, lack of market linkages, uncertainties in weather, and most importantly a serious lack of technological breakthrough. The current intervention aims to bridge many of these gaps. In this chapter, we present an innovative technological intervention in agriculture for the state of Andhra Pradesh (AP).

12.2 OVERVIEW: AGRICULTURE SECTOR IN ANDHRA PRADESH

With a contribution of more than 29% of the gross state domestic product (GSDP) from agriculture and allied sectors, AP can be described as an agro-based economy. While there are 165 crops enumerated in the state, 10 major crops occupy 80% of the gross cropped area. The primary sector

DOI: 10.1201/9781003487531-16

(activities that harness natural resources such as fishery, agriculture, horticulture, etc.) in AP provides employment to 46 lakh farm families, or around 62% of the state's population. Thus, the performance of the sector determines the progress of the state. Recognizing the importance of the primary sector, the state introduced a separate annual budget for agriculture from 2014–15.

There were several challenges which were slowing down the growth of the primary sector in the state like frequent damages due to natural calamities, poor soil health due to indiscriminate use of chemical fertilizers, increasing production costs, and inefficient market linkages.

Realizing the importance of agriculture to the state's economic health and the well-being of its citizens, the government of Andhra Pradesh (GoAP) has initiated several programs to uplift the state's agricultural sector. With the advent and advancement of satellite data technologies, sensors, big data, artificial intelligence, and geospatial and other modern computer science technologies, digital technologies are helping improve agriculture's productivity and leveraging digital technologies to optimize the primary sector growth in the state in one step in that direction.

The government initiated several applications to enhance efficiency at all levels and reach out to farmers with timely advisories, like digitizing land records, applications for crop booking, and delivery of extension services. However, there is still scope to bridge the gap in bringing together the data from different functions, establishing a comprehensive farmer-centric approach to delivery of services, and authenticating the technological insights through proper ground validation.

By leveraging technology to improve delivery, a list of services offered in agriculture sector is provided in Figure 12.1. The solution has been built to help not only farmers but also various stakeholders involved in agriculture and allied activities in the state of Andhra Pradesh. Overall, the solution provides support for all the activities and services that the farmers need to access during the entire farming lifecycle.

Generally, many state and central government schemes are available for farmers. However, many eligible persons cannot make use of these schemes due to a lack of knowledge or awareness. The goal of the solution is to provide access to all the relevant persons regarding various government schemes in a timely and transparent manner. This is done in many ways and at every step of the agricultural cycle.

Planning the next crop to be sown is often not entirely straightforward, and in the absence of any scientific planning tool, the farmers often must rely on their gut feeling or word of mouth from others in the community. The solution provides detailed planning for crops sown based on historical data, current climate conditions, and market demand.

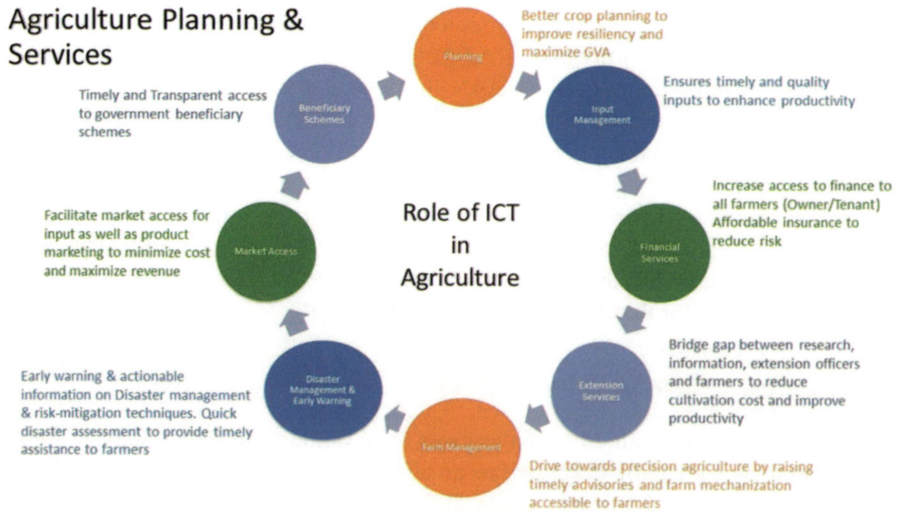

FIGURE 12.1 Snapshot of the services offered in the agriculture sector in Andhra Pradesh.

To execute the plan, the farmer also needs insights on the specific inputs to be provided based on the soil type, climate conditions and crop to be grown, which this solution offers. The solution does not stop at advising on crop planning or choice of input. Still, it continuously provides farm management and risk mitigation advisory services based on continuous monitoring of the crop, which will help the farmer act in real time during natural disasters and other instances. The solution aims to increase farmer productivity and reduce farming costs.

12.3 DECISION SUPPORT SYSTEM

Our team conceptualized and implemented the Andhra Pradesh Agriculture Information and Management System (APAIMS). We aimed to implement synergistic data models to gain evidence-based insights across several modules (Figure 12.2). The following is the solution overview, the conceptual models, and the description for some of the modules.

APAIMS is an innovative, integrated, and user-friendly integrated digital solution for GoAP, consisting of digital infrastructure, system software, databases, data analysis, and dashboards consisting of management information system (MIS) and geographic information system (GIS) views. The dynamic platform is hosted on the cloud to ensure elasticity and scalability.

The platform can analyze various data sets to produce a meaningful advisory. The digital platform envisages quality delivery of advisory and support services to the main stakeholder, the farmer. It has been designed to support the feeding in of real-time information from on-ground stakeholders/functionaries. The integrated platform captures information such as soil health card data, localized weather data from automatic weather stations (AWSs), and soil moisture data from sensors. It superimposes all this information with satellite data-derived insights, information on water availability, geospatial data, crop phenology models, and so on. The APAIMS digital platform capabilities include the assimilation of complex data backed by analytics and a knowledge engine to produce holistic "farm-specific" advisories.

This integrated digital platform and agriculture information and management system help in multiple areas, like bringing all the data sources in silos with different government departments and third-party systems into one unified big data database and digital platform. Further, they also provide a means to base decisions on real-time data, including weather. They bring scientific models

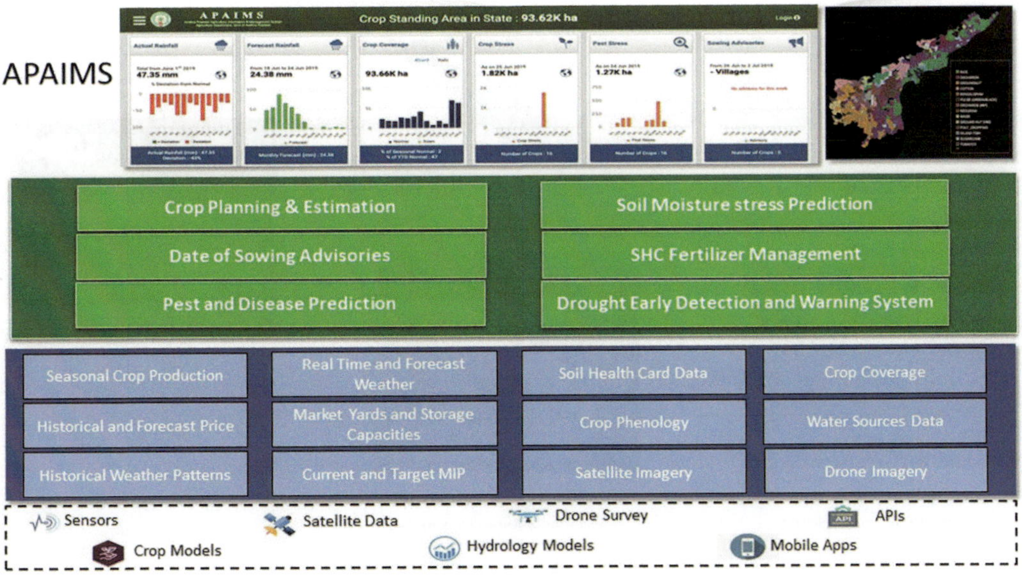

FIGURE 12.2 Overview of the Andhra Pradesh Agriculture Information and Management System.

and validated data together to bring actionable advisories. Finally, they provide a digital mechanism (app, SMS) to send specific and customized advisories to each of the state's agri-extension officers and farmers. Some modules of APAIMS are described in the following sections.

12.4 AGRO-CLIMATIC CROP ZONE PLANNING

This module aims to help the department plan the cropping pattern based on several parameters like crop-specific water requirement, water availability, and soil type to arrive at the optimum combination of crops suitable to the area so as to enhance the overall GVA. The following depicts the agro-climatic crop zone (ACZ) planning implemented in APAIMS.

For example, in AP, data shows that paddy is grown in areas with limited rainfall, thus depleting the groundwater levels; groundnut is grown in areas with less than 300 mm seasonal rainfall, as

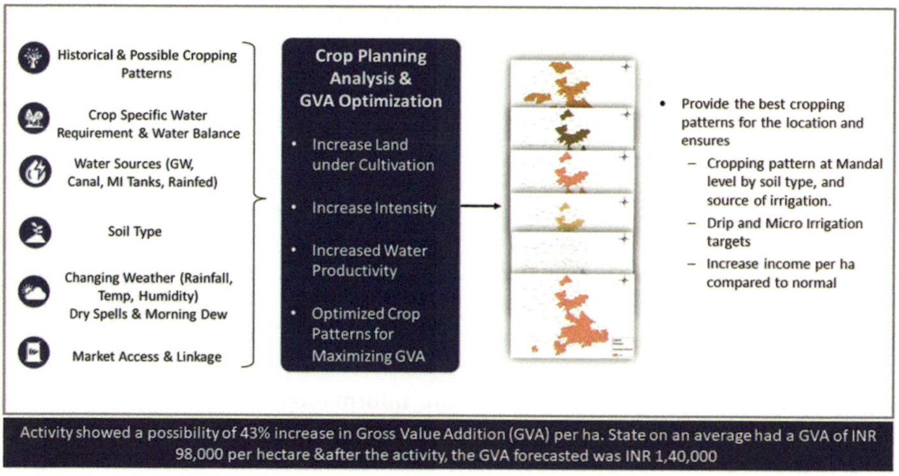

FIGURE 12.3 Module on agro-climatic crop zone planning.

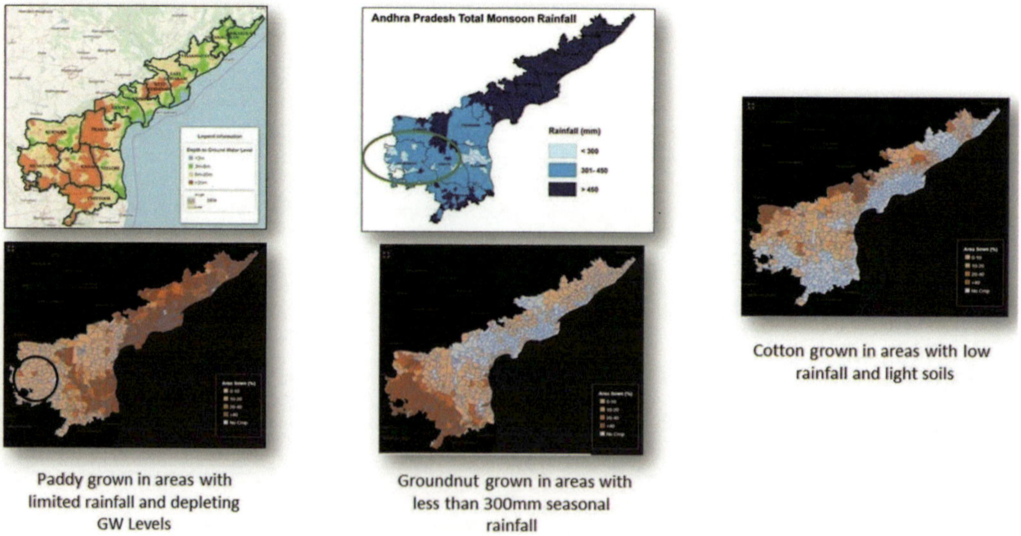

FIGURE 12.4 Requirements for agro-climatic crop zone planning in Andhra Pradesh.

shown. Based on the ACZ analysis, there is a huge scope to bring in evidence-based insights to the department on the cropping pattern in order the optimize the overall GVA.

12.5 DATE OF SOWING ADVISORY

Like insights on what to sow, it is also possible to reap significant benefits from advising when to sow based on weather parameters, historical data, and so on, as shown in the following. The date of sowing advisory is a crucial component of agricultural planning and management, especially for rainfed crops. It provides farmers with recommendations regarding the optimal timing for sowing crops based on local climate, soil conditions, crop varieties, and market demand. This advisory is typically generated by agricultural experts or government agencies using historical weather data, predictive modeling, and agronomic knowledge. By following the recommended sowing dates, farmers can maximize crop yields, minimize risks associated with adverse weather conditions, and enhance overall farm productivity. Additionally, the date of sowing advisory can play a vital role in sustainable agriculture practices by promoting efficient resource utilization and reducing environmental impact.

The following graph shows the crop's success based on when the crop is sown. The vertical axis plots a composite score based on various factors against weekly progress. An example is the groundnut crop in the Anantapur district of AP. The advisory considers weather parameters, historical weather data, crop stages, and water requirements, and so on.

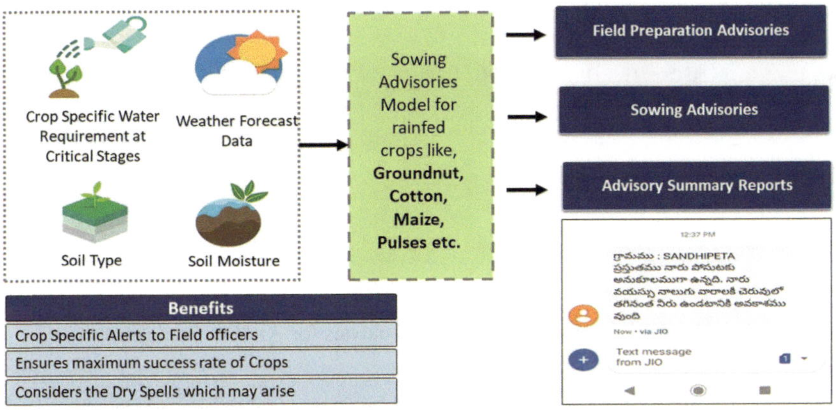

FIGURE 12.5 Date of sowing advisory for farmers in Andhra Pradesh.

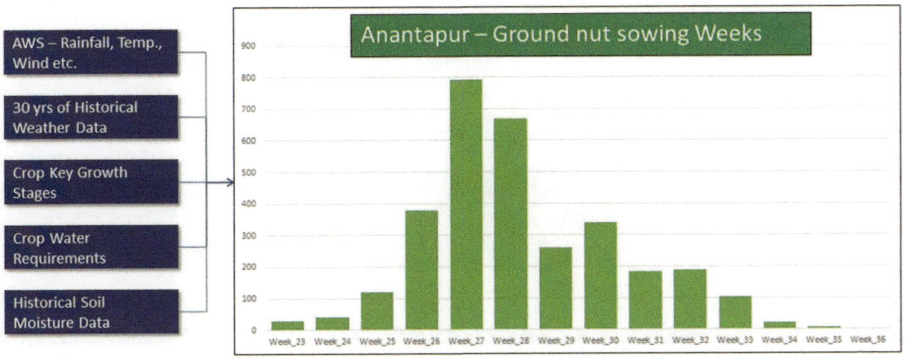

FIGURE 12.6 Representation of the crop's success factor based on the time of sowing.

12.6 CROPS SOWN ANALYSIS

The solution has an integrated, user-friendly dashboard with the historical data of all the crops grown across the state from 2015 until the present year. The module displays progressive sowings for the current year as well. The crops sown analysis would give the user insights on location-wise dominant crop (in GIS), total land utilization, location-wise total crops sown area, crop-wise total sown area, and so on.

Statistical and spatial distribution of crops sown analysis is provided in Figures 12.7 and 12.8. The figure shows the crops sown acreage based on the Department of Economics and Statistics (DES) data in Andhra Pradesh for both the kharif and rabi seasons. The analysis is segregated into areas like agriculture, horticulture, and aquaculture. Further, a crop-wise analysis can be carried out using the dashboard. The dashboard also shows the district-wise analysis of the crops sown and the area under crop. Geospatial information system representation of the crops sown across various geographies like district, mandal, and village are showcased in Figure 12.8.

FIGURE 12.7 Statistical representation of crops sown analysis for Andhra Pradesh.

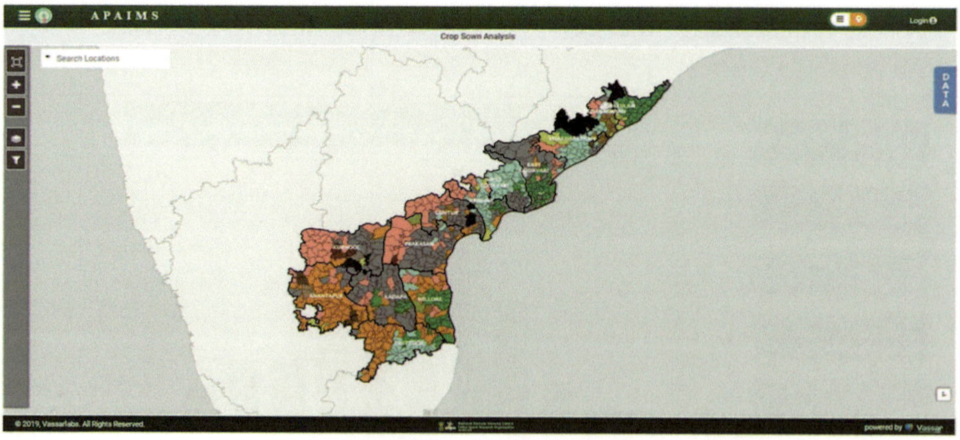

FIGURE 12.8 GIS representation of the crops sown analysis for Andhra Pradesh.

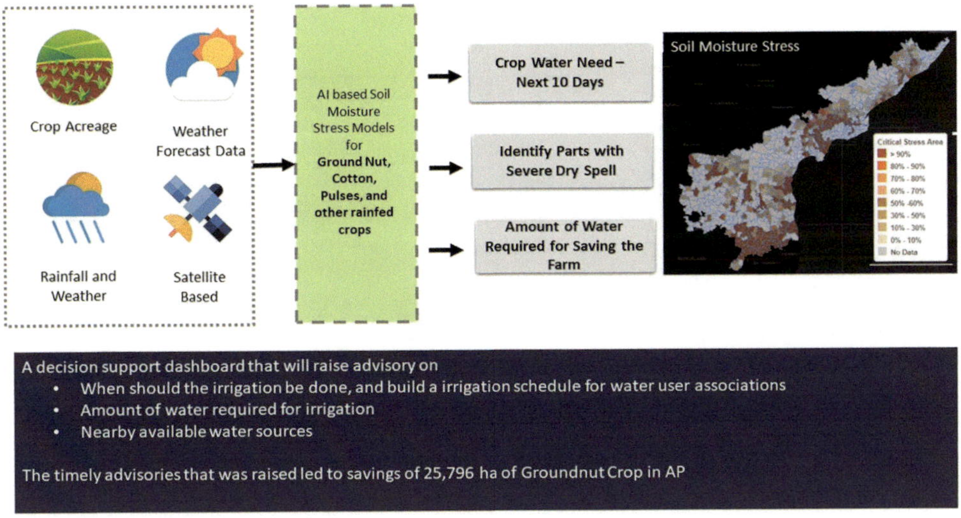

FIGURE 12.9 Soil moisture stress module for APAIMS application, Andhra Pradesh.

12.7 SOIL MOISTURE STRESS ADVISORIES

Soil moisture stress refers to the soil lacking adequate moisture to support optimal crop growth and development. It can occur due to various factors, such as insufficient rainfall, high temperatures, or poor irrigation practices. Soil moisture stress negatively impacts crops by hindering germination, reducing nutrient uptake, and impairing overall plant health. Monitoring soil moisture levels is crucial for farmers to implement timely irrigation strategies and mitigate the effects of moisture stress. By maintaining optimal soil moisture conditions, farmers can promote healthy crop growth, improve yield potential, and enhance water use efficiency in agriculture. Addressing soil moisture stress also contributes to sustainable farming practices by conserving water resources and minimizing environmental impact.

This module provides insights into whether a crop is under critical stress (for the next 7-day window) by considering the soil moisture and rainfall forecast. This is primarily useful for effective irrigation management of rainfed crops. The conceptual model for the module is shown in Figure 12.9.

12.8 PEST/DISEASE STRESS ADVISORIES

Pest and disease stress advisories are essential tools in agricultural management to mitigate the impact of pests and diseases on crop production. These advisories provide farmers with timely information on their region's prevalence, distribution, and severity of pest and disease outbreaks. They often include recommendations for integrated pest management strategies, such as cultural practices, biological control methods, and judicious use of pesticides. By following these advisories, farmers can implement proactive measures to prevent or minimize pest and disease damage to their crops, thereby reducing yield losses and preserving crop quality. Additionally, pest and disease forewarning and advisories contribute to sustainable agriculture by promoting environmentally friendly pest control practices and reducing reliance on chemical pesticides.

This module aims to raise predictive alerts for pest/disease attacks based on weather-based rules and historical data to prepare the farmers for preventive/remedial actions like spraying and others specific to crops and pests/diseases. The dashboard for this module also displays the pest/disease rules for the selected crop. Figure 12.10 shows the conceptual model of this module.

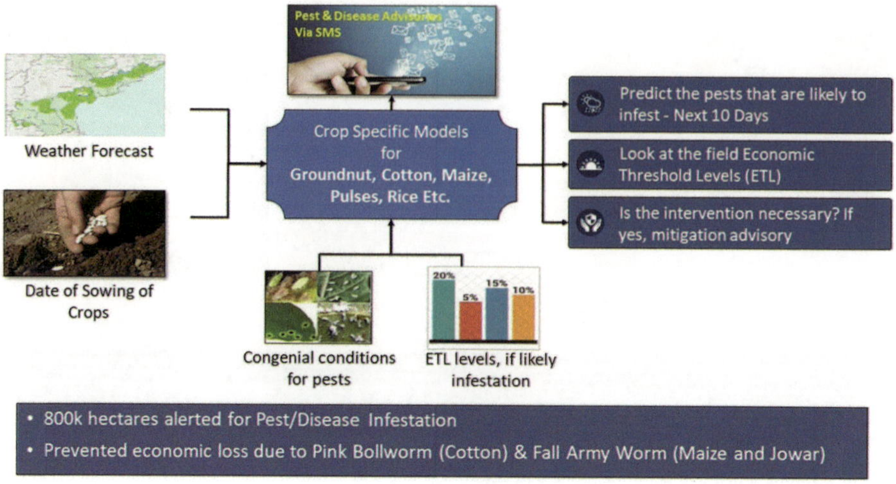

FIGURE 12.10 Pest or disease stress advisory for APIMS application of Andhra Pradesh.

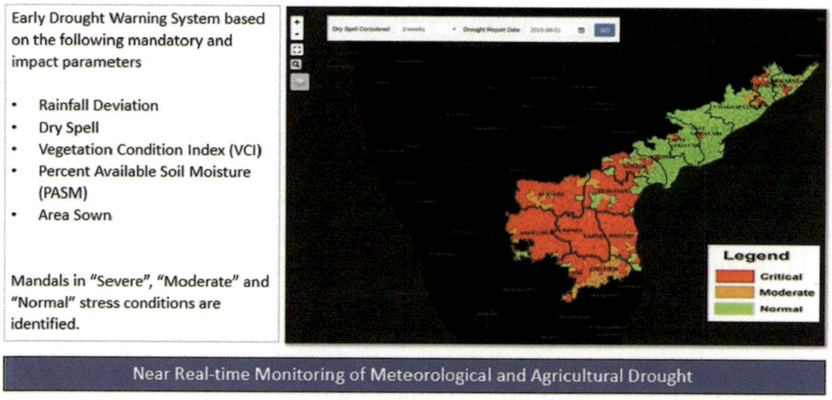

FIGURE 12.11 Early drought warning system in APAIMS application of Andhra Pradesh.

12.9 EARLY DROUGHT WARNING SYSTEM

The early drought warning system (EDWS) plays a crucial role in agriculture by providing timely and accurate information about impending drought conditions. By monitoring various meteorological parameters such as rainfall, temperature, humidity, and soil moisture levels, the EDWS can identify early signs of drought onset and assess its severity. This information enables farmers, policymakers, and other stakeholders to take proactive measures to mitigate the impacts of drought on agriculture.

This module in APAIMS presents a real-time view of the drought status across the state, per the revised Manual for Drought Management published by the Ministry of Agriculture, Cooperation and Farmers Welfare, GoI, India. This helps the state government in declaring drought at the right time to secure benefits for its farmers.

In addition to the mentioned modules, the APAIMS solution provides insights on soil health-based recommendations, crop production analysis, farm mechanization analysis, and crop health indicators. All in all, a summary view of the APAIMS solution can be seen in Figure 12.12.

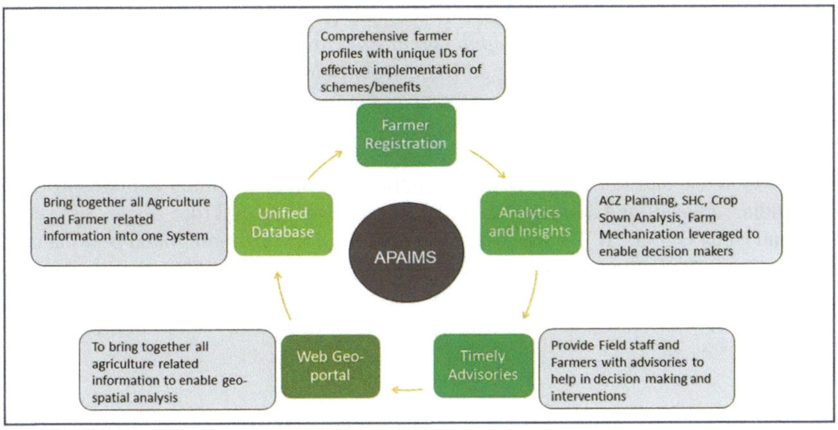

FIGURE 12.12 Summary view of APAIMS application in Andhra Pradesh.

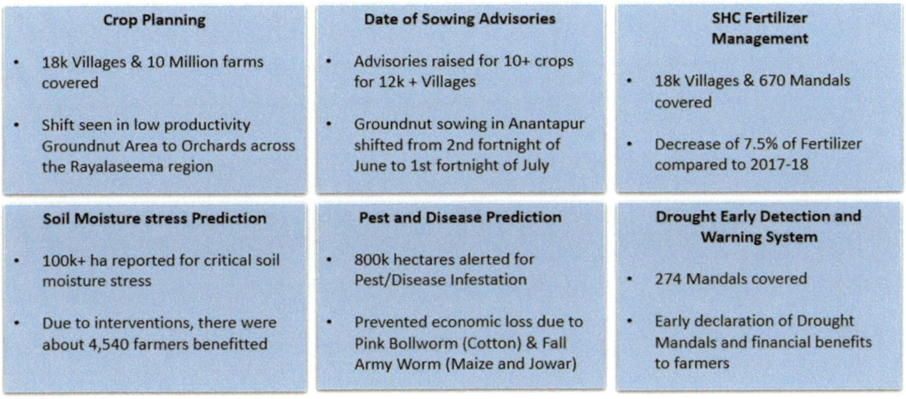

FIGURE 12.13 Benefits of the APAIMS application of Andhra Pradesh.

12.10 IMPACT OF APAIMS

The coverage and impact of the APAIMS solution in the state of Andhra Pradesh in the year 2018–19 is summarized in the following. It saw remarkable success by providing timely alerts, planning, and better management.

Overall, the solution helped 18,000 villages and 10 million farms. Proper advisories were raised for over 10,000 crops across 12,000 villages. Again, 18,000 villages and 670 mandals were provided with fertilizer management, which improved crops.

About 4540 farmers benefited from critical soil moisture stress reporting done by the solution. Further, 8,00,000 hectares of land were alerted beforehand regarding pest and disease infestation. A drought warning was provided across 274 mandals, and financial benefits were provided to eligible farmers in advance to save their crops and livelihoods. The benefits of the APAIMS system are highlighted in Figure 12.13.

BIBLIOGRAPHY

Chaubey, Dayanidhi, Ved Prakash, Arvind Bhai Patel, and Tikam Chand Yadav. "Role of agro-meteological advisory services on risk mitigation in agriculture." *Indian Journal of Pure & Applied Biosciences* 6 (2018): 27–32.

Kukar, Matjaž, Petar Vračar, Domen Košir, Darko Pevec, and Zoran Bosnić. "AgroDSS: A decision support system for agriculture and farming." *Computers and Electronics in Agriculture* 161 (2019): 260–271.

Mir, S. A., M. Qasim, Y. Arfat, T. Mubarak, Z. A. Bhat, J. A. Bhat, S. A. Bangroo, and T. A. Sofi. "Decision support systems in a global agricultural perspective-a comprehensive review." *International Journal of Agriculture Sciences* 7, no. 1 (2015): 403–415.

Singh, K. K., Kripan Ghosh, S. C. Bhan, Priyanka Singh, Lata Vishnoi, R. Balasubramanian, S. D. Attri, Sheshakumar Goroshi, and R. Singh. "Decision support system for digitally climate informed services to farmers in India." *Journal of Agrometeorology* 25, no. 2 (2023): 205–214.

Singh, Niranjan, and Neha Gupta. "ICT based decision support systems for integrated pest management (IPM) in India: A review." *Agricultural Reviews* 37, no. 4 (2016).

Zhai, Zhaoyu, José Fernán Martínez, Victoria Beltran, and Néstor Lucas Martínez. "Decision support systems for agriculture 4.0: Survey and challenges." *Computers and Electronics in Agriculture* 170 (2020): 105256.

13 Use of Satellite Imagery and Analytics for Replacing Crop-Cutting Experiments for Agricultural Yields

Krishna Kumar

ABSTRACT

The crop cutting experiment (CCE) is an efficient assessment method used to determine the estimated yield that can be expected in that agricultural cycle. Earlier methods consisted of random sampling of areas where various soil and crop cycle factors were tested to indicate harvest criteria over a larger area. With the help of artificial intelligence and machine learning, digital solutions have been introduced in CCE to help bring focused results.

These digital solutions, SmartFarm and SmartRisk, use ground-level data and satellite imagery to identify the plots that are appropriate for these experiments. A dedicated and highly skilled data science team analyses millions of data points. It runs them through numerous criteria to zero in on farm plots that will provide the most accurate sample for the region. In this way, ambiguity is removed from plot sampling, making crop insights value rich and focused.

13.1 INTRODUCTION

Every day, we wake up to a new world. An ever-evolving, success-driven, and overly competitive world. It all starts with an idea for a company to bring a dream to reality. What makes a company truly successful? CropIn's journey has always been about making farmers' lives better than yesterday – with nearly half of India's workforce engaged in the agriculture sector, this is crucial. The impact that CropIn has had, owing to its AI and ML-powered smart and sustainable solutions, has been tremendous in terms of improving farmer livelihoods. Since its inception, CropIn has designed its solutions, considering the challenges farmers face globally and how technology can bring about much-needed change for them. CropIn has impacted more than 2.1 million farmers and digitized 6.1 million acres of farmland with projects in 52+ countries and 225+ happy clients.

The crop cutting experiment (CCE) is a field-based method used to estimate crop yield and production. It involves randomly selecting sample plots within a field, harvesting and weighing the crops from these plots, and using statistical analysis to extrapolate the results to estimate the yield for the entire field or region. CCEs are typically conducted for major crops such as rice, wheat, maize, and pulses.

The importance of crop cutting experiments in agriculture lies in its role in providing accurate and reliable estimates of crop yield, which are essential for various purposes:

- **Yield Estimation**: CCEs provide crucial data for estimating crop yields at the field, village, district, and national levels. These estimates are used by policymakers, researchers, and agricultural stakeholders for planning, policymaking, and resource allocation.

DOI: 10.1201/9781003487531-17

- **Crop Insurance**: Crop yield estimates from CCEs are used to assess crop losses and determine compensation payouts under crop insurance schemes. Accurate yield data ensures fair and timely compensation for farmers affected by crop failures due to natural disasters or adverse weather conditions.
- **Agricultural Research**: Researchers use CCE data to study crop performance, evaluate new crop varieties, assess the impact of agricultural practices and technologies, and identify factors influencing crop yields. This information helps improve agricultural productivity and resilience to various stressors.
- **Food Security Planning**: CCEs contribute to food security planning by providing reliable information on crop production levels and potential shortages. Governments use this data to formulate policies for managing food supplies, imports, exports, and distribution to ensure adequate food availability for the population.
- **Policy Formulation**: Policymakers rely on CCE results to design and implement agricultural policies, programs, and interventions aimed at promoting sustainable agriculture, enhancing farmer income, and addressing food security challenges. Accurate yield estimates help policymakers make informed decisions to support agricultural development and welfare.

Overall, crop cutting experiments play a vital role in agricultural monitoring, planning, and decision-making processes, contributing to the sustainable management of agricultural resources and the welfare of farming communities. The crop cutting experiment (CCE) is an efficient assessment method used to determine the estimated yield that can be expected in that agricultural cycle. Earlier methods consisted of random sampling of areas where various soil and crop cycle factors were tested to indicate harvest criteria over a larger area.

With the help of artificial intelligence and machine learning, digital solutions have been introduced in CCE to help bring focused results. These digital solutions, SmartFarm and SmartRisk, use ground-level data and satellite imagery to identify the plots that are appropriate for these experiments. A dedicated and highly skilled data science team analyses millions of data points. It runs them through numerous criteria to zero in on farm plots that will provide the most accurate sample for the region. In this way, ambiguity is removed from plot sampling, and this makes crop insights value rich and focused.

13.2 TRADITIONAL METHODS FOR CROP CUTTING EXPERIMENT

The traditional methods followed in farming bring a lot of inaccuracies in data collection, and human error is a factor that cannot be avoided with these methods. Owing to this erroneous and cumbersome data-gathering process, farmers often faced delays in the disbursement of insurance claims they raised due to the lack of a proper assessment method. To change this, the government of India adopted a very efficient data collection and analysis method called the crop cutting experiment. This was implemented with the help of a strategic partnership between the government of India and CropIn. The crop cutting experiment is done every year by the Indian government as an assessment method that can be used to determine the estimated yield that can be expected in that agricultural cycle. The process had to ensure data collection and analysis were error free and accurate and ensure proper settlement of farmer insurance claims.

The traditional method of the CCE is based on the yield component method, where sample locations are selected based on a random sampling of the total area under study. Once the plots are selected, the produce from a section of these plots is collected and analyzed for several parameters such as biomass weight, grain weight, moisture, and other indicative factors. The data gathered from this study is extrapolated to the entire region and provides a fairly accurate assessment of the average yield of the state or region under study.

13.2.1 Introducing Technology in Crop Cutting Experiments

The use of technology in agriculture has made the practice of farming a lot more predictable and efficient. Compared to the traditional method of CCE using random sampling, remote sensing and other technological advancements in agriculture provide a far more accurate and timely estimation of yield. But the impact of technology starts much earlier. In addition to conducting CCEs to estimate the total yield, the central government uses this data for its Pradhan Mantri Fasal Bima Yojana (PMFBY), which is a scheme that helps insurance companies disburse payment for farmers' insurance claims seamlessly. The PMFBY requires each state to carry out at least four CCEs in every village panchayat for each crop and submit the yield data to insurance companies within one month of harvest.

Pradhan Mantri Fasal Bima Yojana is a crop insurance scheme launched by the government of India to provide financial support to farmers in case of crop loss or damage due to natural calamities, pests, diseases, or adverse weather conditions. Under the PMFBY, farmers pay a nominal premium, and the remaining premium is subsidized by the central and state governments. The scheme aims to ensure maximum coverage of crops and farmers, promote sustainable agricultural practices, and mitigate the financial risks faced by farmers due to crop failure. PMFBY offers comprehensive coverage for all stages of crop production, including pre-sowing, sowing, and post-harvest losses. It also provides coverage for localized calamities such as hailstorms and landslides. The scheme employs technology-driven processes for quicker claim settlement, transparency, and efficiency. PMFBY has significantly increased the coverage of farmers under crop insurance and has helped in reducing the financial burden on farmers during crop losses.

The data gathered from CCE is useful to multiple stakeholders in the agricultural value chain. While governments use it for planning agricultural policies and programs for the future, the information helps financial institutions with all the inputs they need before offering loans or insurance coverage if there is a poor harvest or crop failure.

13.2.2 Role of CropIn in Technology Implementation

The Mahalanobis National Crop Forecast Centre (MNCFC) has been established under the Department of Agriculture, Cooperation and Farmers' Welfare, Ministry of Agriculture & Farmers' Welfare, Government of India, for applications of space and geospatial technology in various domains of agriculture. MNCFC was established initially to provide in-season crop forecasts and assessment of drought situations using state-of-the-art techniques and methodologies developed by the Indian Space Research Organization (ISRO).

CropIn is a chosen partner by the Mahalanobis National Crop Forecast Centre of the Department of Agriculture (Government of India) to provide technology for plot and farm-level crop stage data to analyze and identify CCE points through smart sampling.

The project was implemented, and the data was collected and analyzed by leveraging CropIn's digital solutions – SmartFarm and SmartRisk. For conducting CCE experiments, CropIn's SmartFarm platform captures the precise location and size of the farm, details of the farmer, and crop details right from the pre-harvest stage. This ensures that the field data is accurate, enabling authorities to use relevant data at the appropriate time easily. SmartFarm is an innovative agricultural application developed by CropIn. It leverages advanced technologies such as artificial intelligence, machine learning, and satellite imagery to provide comprehensive farm management solutions to farmers and agribusinesses. The SmartFarm application offers various features including crop monitoring, pest and disease detection, weather forecasting, soil health analysis, and yield prediction. It enables farmers to make data-driven decisions by providing real-time insights into their farm operations. Through SmartFarm, farmers can optimize resource utilization, improve crop productivity, reduce input costs, and mitigate risks associated with crop cultivation. The application is user friendly and accessible via smartphones, making it convenient for farmers to manage their farms efficiently.

Overall, SmartFarm contributes to sustainable agriculture practices and empowers farmers with actionable information to enhance their agricultural productivity and profitability.

CropIn's solution, SmartRisk, an AI and ML-powered digital platform, uses ground-level data and satellite imagery to identify the plots that are appropriate for these experiments. A dedicated and highly skilled data science team analyses millions of data points. It runs them through numerous criteria to zero in on farm plots that will provide the most accurate sample for the region. With the help of this data, authorities can quickly identify the suitable plots to be included in the study, removing all ambiguity from the selection process.

The benefits of technologically-aided CCE are not just limited to choosing suitable samples for study. Apart from offering a more optimized method of plot selection, many of these products also help provide the government and insurance companies with scientific, scalable, and accurate reports for future processing. The inclusion of technological intervention addressed the issues of many CCEs being conducted with limited humanpower during a short harvesting window.

SmartRisk enables MNCFC with the following capabilities:

- Identification of crop stages as well as crop yield forecasts at a regional level
- Identification of CCE points at the plot/farm level
- Increased visibility of CCE progress with near real-time insights

Plot-level data insights further help in the following activities:

- Crop detection and health monitoring
- Identification of crop harvest to efficiently conduct CCE
- Monitoring CCE progress through a one-of-a-kind comprehensive dashboard
- Analyzing the sowing cycle and estimating the harvesting window

13.2.3 Impact of Technology Implementation

Technology helps address the issue of conducting many CCEs with limited manpower during a short harvesting window. Using a digital platform such as CropIn considerably reduces the paperwork and subsequent possibilities of human error. Additionally, CropIn provides trained field managers to oversee the process and gather the required data, thus reducing the burden on the government's side. With the help of these scientific methods, the government can improve overall efficiency by utilizing its resources in the best possible way.

For the insurance industry, the data provides a more accurate report of the crop in question and allows for timely claim settlement. The reports derived using CropIn's platform are data-driven and precise, hence removing possibilities of fraudulent claims or inaccurate payment disbursals. It also allows insurance companies to customize crop insurance schemes and products based on real-time data gathered from each farm. For the farmers, digitizing the CCE process enables impartial settlement of claims, which immensely benefits farmers. It reduces the stress on the farmer to provide proof of his/her claims, thereby reducing the effort and time spent in the process.

13.2.4 How Does the Solution Work?

The CropIn system captures total and near real-time data across the supply chain so that the management can monitor processes on a single interactive dashboard. The process is shown in Figure 13.1.

The platform is also used as a digital platform to manage the entire seed certification process – capturing data from registration, inspection, and post-harvest stages (Figure 13.2).

The platform leads to a smart farm setup. CropIn's SmartFarm platform provides comprehensive modules to set up the complete program at the cluster level and configure crops, their input

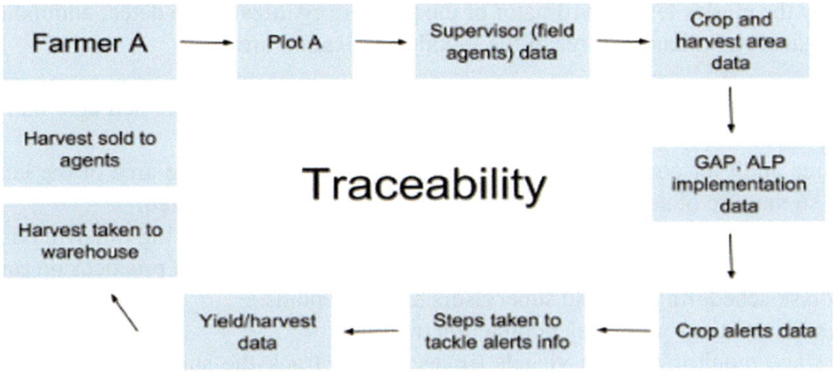

FIGURE 13.1 The traceability of the information collected by the application.

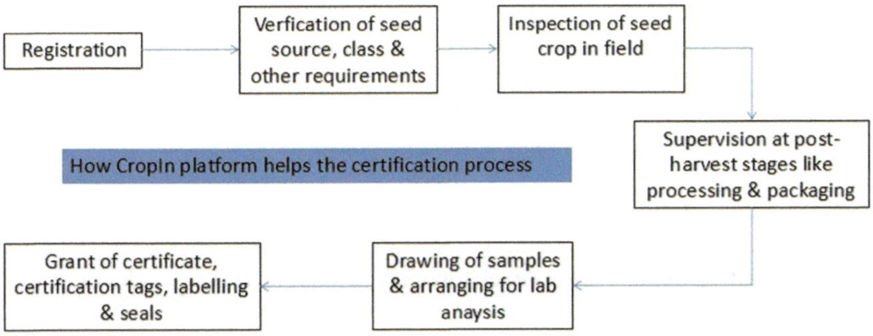

FIGURE 13.2 The verification process for the data capture in the application.

recommendations and package of practices, alerts, and advisory, and harvest estimates. The application creates a user hierarchy and assigns responsibilities based on user roles. Adding and editing details like farmer, crop, activity, and harvest from the web and other sources is applicable at any point during the season. The solution involves the following aspects:

- Comprehensive map-based dashboard to give an overall perspective of how the clusters perform. The farmer's spread can be seen over the map with the complete status of their crops, pest and disease incidence, harvest, and activities so that real-time action can be taken to help farmers.
- The reports section provides operational day-to-day and consolidated figures in the fields of farmer management, crop management, agri-input management, field team management, and harvest management.
- Alerts and advisories to provide real-time guidance on field requests raised through the module with a proper evaluation on the impact on the crop with timely advisories.
- Comprehensive weather intelligence and farmer advisories.
- Language: The entire web application has the option to be in Hindi, English, or regional languages depending on the user's choice.

The smart farm is further integrated with mobile applications. The mobile application is provided to representatives for real-time data capturing and reporting. Supervision is done through the mobile application by the cluster-level coordinator of the partner organization to detect anomalies and take corrective measures. Salient features of the mobile application are:

- Farmer member registration along with farmer photo, crop type, acreage, and sowing information.
- Plot area audit for correct acreage measurement – To measure the area of the land accurately so that use of seeds, fertilizer, and pesticides can be optimized.
- Farm activities and package of practice (POP) management ensuring timely advisories to farmers based on their crop types and ensuring completion of best practices on time.
- Daily task scheduling for field supervisors based on plans.
- Field staff location-based task completion tracking.
- Crop stage monitoring with visuals (images) – To track the stage of the crop and give adequate information about crop status.
- Advisory module for pest and disease resolution from a central location.
- Weather updates and crop risk.
- Harvest date re-estimation – To correct forecasts based on crop performance.
- Fertilizer and pesticide recommendations – Provide guidelines per standard POPs.
- Harvest details and consignment building – To optimize the supply chain and reduce loss.
- Training – To provide training and capture details.

Finally, the system was connected to a market linkage platform.

- The market linkage platform is integrated to the data platform with the ability to showcase farms' data, farmer member's data, and crop stage data, which buyers can access globally.
- The platform showcases a cluster-level forecast to buyers every week for the next four weeks for buyers to engage proactively.
- Buyers can reach out directly to the representative and carry out negotiations and transactions.

13.3 IMPACT OF THE APPLICATION ON PRADHAN MANTRI FASAL BIMA YOJANA

The data generated through CCE is utilized by the Pradhan Mantri Fasal Bima Yojana, the crop insurance program for India. The data assists the insurance companies for various causes, including seamlessly disbursing payment to farmers' insurance claims, and planning agricultural policies. Further, financial institutions use this data to assess loans and insurance coverage in the case of a poor harvest.

Indian agriculture is mainly composed of smallholdings. Up to 80% of the farmers in India are smallholders and use traditional methods of cultivation and farm management. Owing to this situation, tracking if enough food is being produced becomes a challenge for the government.

The biggest drawback of this traditional method for CCE is that it depends on several variables such as administrative setup, type and size of the field staff, farmer cooperation, and harvest conditions. Especially in a scenario where there are nearly 2.5 lakh gram panchayats in India that are scattered, along with inadequately trained human labor or the time to facilitate these experiments effectively, there needs to be a more efficient way of utilizing resources and obtaining an accurate yield estimation within a short harvesting window.

CropIn's digital solutions, SmartFarm and SmartRisk, were implemented through the PMFBY to disrupt the process to increase accuracy of results and cut down on the various challenges in traditional data capturing. As mentioned, SmartFarm captures details of the farm and the farmer right from the pre-harvest stage, creating accurate field data for authorities.

The inclusion of technological intervention addressed the issues of many CCEs being conducted during a short harvesting window with limited human power. This solution helped in achieving the objective with limited resources and more accuracy, resulting in the timely settlement of claims.

BIBLIOGRAPHY

Lobell, David B., George Azzari, Marshall Burke, Sydney Gourlay, Zhenong Jin, Talip Kilic, and Siobhan Murray. "Eyes in the sky, boots on the ground: Assessing satellite-and ground-based approaches to crop yield measurement and analysis." *American Journal of Agricultural Economics* (2019): aaz051.

Peltonen-Sainio, Pirjo, Lauri Jauhiainen, Eija Honkavaara, Samantha Wittke, Mika Karjalainen, and Eetu Puttonen. "Pre-crop values from satellite images for various previous and subsequent crop combinations." *Frontiers in Plant Science* 10 (2019): 462.

Ranjan, Avinash Kumar, and Bikash Ranjan Parida. "Predicting paddy yield at spatial scale using optical and synthetic aperture radar (SAR) based satellite data in conjunction with field-based crop cutting experiment (CCE) data." *International Journal of Remote Sensing* 42, no. 6 (2021): 2046–2071.

14 Harnessing Technology to Create Smart Farms and Achieve Agricultural Progress

Case Studies across India

Praveen Pankajakshan and Krishna Kumar

ABSTRACT

Three separate projects, the Jeevika World Bank Project, Mission 2020: Lakhpati Kisan – Smart Villages Project, and Seed Potato Traceability Project, were launched across various states of India where technology was introduced in farming methods to increase livelihoods, issue farm advisories, counter climate change, uphold the lives of tribal households, and improve seed quality.

Farm data is captured at critical points to establish end-to-end traceability to the life cycle of crops, starting from pre-harvest until final harvest. Raw data is processed, and farmers are provided with data insights, including weather alerts, pest forecasts, management practices, monitoring production and harvest, and adhering to export compliance. Using the technology, 80% of farm households implemented the advisories issued, 92% demonstrated climate-resilient agricultural practices, and 90% declared satisfaction with the satellite-based advisory.

14.1 INTRODUCTION

SmartFarm is an innovative agricultural application developed by CropIn. It leverages advanced technologies such as artificial intelligence, machine learning, and satellite imagery to provide comprehensive farm management solutions to farmers and agribusinesses. The SmartFarm application offers various features, including crop monitoring, pest and disease detection, weather forecasting, soil health analysis, and yield prediction. It enables farmers to make data-driven decisions by providing real-time insights into their farm operations. Through SmartFarm, farmers can optimize resource utilization, improve crop productivity, reduce input costs, and mitigate risks associated with crop cultivation. The application is user friendly and accessible via smartphones, making it convenient for farmers to manage their farms efficiently. Overall, SmartFarm contributes to sustainable agriculture practices and empowers farmers with actionable information to enhance their agricultural productivity and profitability.

Further details of SmartFarm are provided in the previous chapter 13. It is a technology solution that uses satellite image data integrated at the farm level and market to create pre-harvest stages, creating accurate field data for authorities.

The inclusion of technological intervention addressed the issues of many crop cutting experiments (CCEs) being conducted during a short harvesting window with limited humanpower. This solution helped in achieving the objective with limited resources and more accuracy, resulting in timely settlement of claims.

We present three case studies where the smart solution was employed.

DOI: 10.1201/9781003487531-18

14.2 THE JEEVIKA WORLD BANK PROJECT

The Sustainable Livelihoods and Adaptation to Climate Change (SLACC) Project for India, in association with the government of India's National Rural Livelihoods Project (NRLP) and supported by the World Bank, aims to empower farmers to adopt climate-resilient practices and adapt to climatic changes and unpredictability.

The JEEViKA World Bank Project,[1] initiated by the government of Bihar in collaboration with the World Bank, aims to improve the livelihoods of rural communities in Bihar, India. JEEViKA, which stands for Jeevika Bihar Rural Livelihoods Promotion Society, focuses on enhancing the economic opportunities for the rural poor, particularly women, through various interventions such as skill development, access to finance, and market linkages. The project prioritizes the empowerment of women by promoting their participation in self-help groups (SHGs) and providing them with training and resources to engage in income-generating activities. Additionally, the project aims to strengthen the institutional capacity of community-based organizations and improve their access to essential services and infrastructure. Through its multifaceted approach, the JEEViKA World Bank Project endeavors to alleviate poverty, promote sustainable livelihoods, and enhance the overall socio-economic development of rural communities in Bihar.

The SLACC project was launched in 2017 in two districts of Bihar, Madhubani and Gaya. The project aimed to devise strategies to help farmers counter climate change's impact on their crops and other farming activities and increase their livelihood. The role of CropIn was as a technical support agency responsible for disseminating weather-based agro-advisories to selected farmers of Gaya and Madhubani. The farmers were spread across four blocks, two from each district, and were provided advisories across three consecutive years, 2017, 2018, and 2019.

The solution designed by CropIn empowered farmers with:

- Weather-based advisories and weather alerts.
- Crop management practices, known as package of practices (PoP), to be undertaken by the farmer for the stage of the crop.
- Pest and disease forecast and management practices.
- Nutrient management practices for the stage of the crop.
- Soil and water management practices to be undertaken for the stage of the crop.
- Animal husbandry.

Using the CropIn platform, the achievements were 80% implementation of a package of practices and advisories, 92% demonstration of climate-resilient agricultural practices, and 90% satisfaction with satellite-based advisories.

14.3 LAKHPATI KISAN – SMART VILLAGES

Lakhpati Kisan is an initiative the government of India launched to transform rural areas and empower farmers to increase their income. The mission aims to achieve the target of having at least one lakh farmer households become "lakhpati" (millionaires) by 2020. Through various interventions and programs, the initiative focuses on promoting sustainable agriculture practices, enhancing productivity, providing access to technology and market linkages, and improving rural infrastructure. It also emphasizes the importance of diversification of income sources, skill development, and capacity building among rural communities. By fostering the development of intelligent villages, the mission aims to create vibrant rural economies, reduce agrarian distress, and improve the overall well-being of farmers and their families across India.

Collectives for Integrated Livelihood Initiatives (CInI) is a philanthropic organization established by Tata Trusts to uplift the lives of central India's tribal households through its Central India

Initiative. In April 2015, CInI launched its ambitious Mission 2020: Lakhpati Kisan – Smart Villages program to bring about 101,000 households irreversibly out of poverty with increased quality of life and life choices and to develop 17 blocks as regional drivers for growth across Maharashtra, Gujarat, Jharkhand, and Odisha. Each of the 101,000 households is envisaged to earn INR 120,000 per annum or more, compared with the baseline average yearly earnings of INR 40,000. Currently, the program impacts over 70,000 households, bringing them in the ambit of institutional structures for driving change.

This initiative focuses on leveraging the best and most advanced technologies in agriculture to address the specific constraints faced by the tribal population in the region. Besides cultivating crops, CInI wants to establish other sources of livelihood, like the production of *lac* and *tussar.*

To achieve the goal of capacity building of the farmers, CInI is primarily leveraging CropIn's SmartFarm as a tool. The company has an active user base of around 40, each handling a minimum of 50 farmers. The objective of CInI was to improve the livelihood of farmers growing vegetables through digitizing and channeling data.

The field staff engages with the farmer daily and collects critical farm data using SmartFarm's easy-to-use modules. This data is stored in CropIn's centralized cloud platform and is readily available to other key stakeholders in the project for effective decision-making. This is their first year of engagement with CropIn, and the field staff is utilizing SmartFarm to monitor capsicum, green peas, watermelon, bottle gourd, cauliflower, chili, cabbage, brinjal, bitter gourd, and tomatoes.

> Achieving aspirations of tribal communities through the focused Lakhpati Kisan programme has been the focus at CInI. For us, innovative technologies and approaches are core for achieving the outcomes of these communities. The partnership with a well-known technology partner such as CropIn is unique and important for us to meet the audacious goals of our programme. We proactively look for this partnership to be taken at scale for larger benefits of small and marginal farmers.
>
> **(Ganesh Neelam, executive director of CinI)**

14.4 PAGREXCO – GOVERNMENT OF PUNJAB PARTNERSHIP ON SEED TRACEABILITY

Punjab produces 27 lakh tonnes of potatoes yearly, so ensuring high seed quality is critical. Given the sub-par, spurious seeds flooding the market and resulting in low yields and poor return on investment, this certification can transform potato seed production. It will also significantly build trust and credibility for Punjab Agri Export Corporation's (PAGREXCO) seeds among potato growers. CropIn has partnered with the Punjab Agri Export Corporation, a government of Punjab enterprise, to leverage technology to improve the quality of potato seed production in the state and incorporate a first-of-its-kind "seed potato traceability" in the entire value chain.

End-to-end traceability for potato seeds is essential to ensure the seed supply chain's quality, safety, and authenticity. Potatoes are a vital crop globally, and the quality of seeds directly influences crop yield, disease resistance, and overall agricultural productivity. End-to-end traceability enables stakeholders to track the journey of potato seeds from production to consumption, including seed selection, cultivation, storage, distribution, and sale. Farmers, seed producers, distributors, and consumers can access crucial information about seed origins, quality standards, handling practices, and certifications by implementing traceability systems. This transparency helps identify and mitigate risks such as counterfeit seeds, contamination, pest infestations, and disease outbreaks. Moreover, traceability facilitates compliance with regulatory requirements, enhances market access, builds consumer trust, and supports sustainable agriculture practices. Overall, end-to-end traceability plays a vital role in safeguarding the potato seed industry's integrity and ensuring the availability of high-quality seeds for farmers worldwide.

CropIn's advanced farm data management solution will empower farmers to capture several parameters through every production stage and monitor them systematically to produce superior-quality seed potatoes. The PAGREXCO field teams will configure the best package of practices for each seed variety on CropIn's platform and will also receive real-time intelligence about local weather, crop growth, crop health, potential pest infestations, and crop diseases. This will help seed-potato growers in Punjab to cultivate superior-quality seeds and improve their efficiency. Finally, CropIn will facilitate two-way communication between PAGREXCO and these farmers, allowing managers to share important plans, advisories, and instruction videos and facilitating farmers' instant access to the PAGREXCO team.

Capturing farm data at critical points establishes end-to-end traceability to the life cycle of seeds. Following tests for quality assurance and certification by PAGREXO, unique QR codes are printed on each seed potato packet during the packaging and warehousing process. This will enable buyers to trace the origin of the seed potatoes and determine precisely where they are coming from, how they were cultivated, and whether they are PAGREXCO certified.

This traceability feature plays a vital role in curbing the sale of counterfeit, low-quality seeds under the name of Punjab-grown seed potatoes. It will also prevent the mixing of high-quality certified seeds with sub-par ones that result in low yields and poor ROI for potato growers. The need for traceability of produce, specifically for marginal and small farm holders, is emerging as an effective pathway for improved access to investment, technologies, knowledge support, inputs, and markets.

PAGREXCO has been asked to intervene in potato seed traceability and certification for the state of Punjab, as there is huge potential in the state. There is a disconnect between the farmers and the certification agencies and a gap in the communication channel between them. A digital platform is required to enable them to achieve complete transparency in the potato seed multiplication programs, ensuring traceability and best certificate-level quality produce.

Technology can be a big enabler to bring transparency in the potato seed production value chain. This is achieved by digitizing the farmer database, giving access to technical know-how on crop planning and management, inputs, credit, post-harvest management, value addition, and better market linkages for farmers.

The CropIn integrated platform under PAGREXCO would digitize all the farmers engaged with them across different districts in the state and make field-level data collection user friendly. The centralized platform would help capture all relevant information through a simple mobile application and give dashboard-based insights on the data captured. Having field-level monitoring capability, all the farm activities are captured in the system, ensuring traceability of the entire value chain.

Some of the challenges made easy by CropIn's system were:

- Supply chain efficiency and traceability
- Ensuring organic and sustainable produce
- Monitoring production and harvest
- Adhering to export compliance

While on the one hand, it boosts credibility for PAGREXCO's brand name and consequently promotes farmers' trust in its products, on the other, it also enables farmers registered with PAGREXCO to fetch a premium price for their certified seeds.

Punjab caters to 60–70% of the entire seed potato demand in the Indian market. Hence, this partnership will impact not only seed-potato farmers in Punjab but also potato growers across India and the international markets. CropIn is excited about partnering with PAGREXCO and being able to improve outcomes at such a large scale.

End-to-end traceability of seeds is pivotal in promoting a sustainable future for agriculture and food systems. By implementing robust traceability systems, stakeholders can track the entire journey of seeds, from production to consumption, ensuring transparency, accountability, and sustainability at every stage. Here's how end-to-end traceability contributes to a sustainable future:

- **Quality Assurance**: Traceability systems allow for the monitoring of seed quality throughout the supply chain, ensuring that only high-quality, certified seeds reach farmers. This helps improve crop yields, reduce waste, and enhance agricultural productivity.

- **Disease Management**: Traceability enables the rapid identification and containment of disease outbreaks by tracking the origins of infected seeds. This proactive approach helps prevent the spread of diseases and minimizes the use of harmful pesticides and chemicals.
- **Environmental Conservation**: By tracing the sources of seeds and inputs, agricultural practices can be optimized to minimize environmental impact. This includes reducing water usage, mitigating soil erosion, and promoting biodiversity conservation.
- **Supply Chain Efficiency**: Traceability systems improve supply chain visibility, enabling better inventory management, logistics planning, and resource allocation. This leads to reduced inefficiencies, lower transportation costs, and improved resource utilization.
- **Compliance and Certification**: Traceability supports compliance with regulatory standards and certifications, such as organic or non-GMO labels. This enhances market access for sustainable products and fosters consumer trust and confidence.
- **Data-Driven Decision-Making**: The data generated through traceability systems can be analyzed to identify trends, patterns, and areas for improvement. This enables evidence-based decision-making and continuous improvement in agricultural practices.

Overall, end-to-end traceability of seeds is a cornerstone of sustainable agriculture, promoting environmental stewardship, economic viability, and social responsibility throughout the seed supply chain. CropIn's solutions transformed the seed production value chain by building technological efficiencies into the long and arduous agricultural life cycle.[2]

NOTES

1 https://projects.worldbank.org/en/projects-operations/project-detail/P090764
2 https://www.cropin.com/blogs/seeding-the-perfect-seeds-for-a-sustainable-food-secure-tomorrow/

BIBLIOGRAPHY

Blaettchen, Philippe, Andre P. Calmon, and Georgina Hall. "Traceability technology adoption in supply chain networks." *Management Science* 71, no. 1 (2025): 83–102.

Hu, Jinyou, Xu Zhang, Liliana Mihaela Moga, and Mihaela Neculita. "Modeling and implementation of the vegetable supply chain traceability system." *Food Control* 30, no. 1 (2013): 341–353.

Paul, Sohini. "Economic empowerment of women: A review of the Jeevika programme." *Development with Justice* (2024): 239–247.

Prakash Anjal, Chhatre Ashwini, Neelam Ganesh, Kumar Sujit, and Duddu Apurva. "Lakhpati Kisan: Transforming agriculture-based livelihood for smallholder farmers in India." Routledge India (2025). doi.org/10.4324/9781003258018.

Siwach, Garima, Sohini Paul, and Thomas de Hoop. "Economies of scale of large-scale international development interventions: Evidence from self-help groups in India." *World Development* 153 (2022): 105839.

15 Supporting Farmers through Comprehensive Real-Time Agricultural Advisories Using Spatial Analytics

Ashish Gautam and Bishakha Bhattacharya

ABSTRACT

Centre for spatial analytics and advanced GIS (C-SAG) focuses on generating reliable ensembled weather products for its agri-GIS and providing advisories to plots and farmers. This chapter talks about the approach C-SAG took in collaboration with IBM to embed necessary weather inputs in the agriculture advisory, which was provided using advanced GIS techniques to 53K farmers in the 2018 season in Odisha across 532 villages, and increase it to 133K farmers.

In the 2019 and 2020 seasons, C-SAG organized a multi-parametric spatial database of about 200-odd parameters village-wise – a rich database for analytics and advanced machine learning processing.

15.1 INTRODUCTION

It is known that data science and all associated technologies are working tirelessly to disrupt and innovate various sectors of industry and markets. That said, one of the biggest big data beneficiaries should be the Indian agricultural sector, as it uses various types of structured and unstructured data, layered on which are artificial intelligence (AI) and machine intelligence (MI) techniques. These interventions range from plot mapping and harvest forecasts to agricultural advisories.

The Centre for Spatial Analytics and Advanced GIS (C-SAG) and Tata Trusts have jointly designed a plot concept for farmer-centric comprehensive agricultural advisories. They have developed a process to build, manage, analyze, and generate real-time GIS-based advisories at the individual plot level. This end-to-end process is being implemented in 532 villages of three districts in south Odisha.

The process involves disaggregating various spatial and non-spatial inputs using complex data analytics methods to a plot level and for a 360° evaluation of a farmer (owner/tiller of the plot). This, in turn, brings integrative spatial analytics support to the farmer.

15.2 CURRENT SCENARIO

C-SAG is an independent entity under the National Institute of Advanced Studies (NIAS) Council of Management (COM) for building deep knowledge capabilities and demonstrating impacting societal benefits in spatial analytics and GIS.

Tata Trusts (TT) works in social welfare and philanthropy and sponsored C-SAG to develop AGRI-GIS – a first-of-its-kind activity that C-SAG designed to impact farmers and their quality of life using advanced spatial analytics and GIS tools. The aim is to help enhance the productivity and incomes of TT beneficiary (and other) farmers by applying spatial analytics and GIS.

DOI: 10.1201/9781003487531-19

To this end, C-SAG developed various knowledge assets for high-end image analytics, advanced meteorological data analytics, crop suitability analytics, indexing analytics for social and economic assessments of farmers, models of agricultural practice assessments, and beneficiary analytics. All of these are developed based on field surveys and market analytics. These are ultimately integrated into a customized advisory for a plot and farmer in real time. In the end, this creates the best advisory currently available to farmers.

At that point, C-SAG generated advisories for 53,000 farmers in 170 villages (of the 532 villages mentioned previously) and will soon extend to 1,33,000 farmers in all 532 villages. Thus, for the first time, C-SAG generated a plot-by-plot customized advisory for each farmer/plot, which is being disseminated and made available at local levels.

15.2.1 Solution

The Agri-GIS engine is a techno-neutral platform. It has been localized and beneficiary oriented. GIS services are administered through a local advisor. The latest technologies of automated farm-level agri-data aggregators, spatial analytics, geo-tagged real-time data, and virtual geographical information have been adopted to create the Agri-GIS.

Weekly multi-spectral and synthetic aperture radar (SAR) images are analyzed in real time to identify paddy crops, rabi season crop status, water body fluctuations, and many other relevant parameters that greatly benefit farmers when notified about in advance. C-SAG has organized a multi-parametric spatial database of about 200-odd parameters village-wise, a rich database for analytics and advanced machine learning processing.

For meteorology analytics, C-SAG accessed forecast and measured data sets from the India Meteorological Department (IMD) and enhanced forecasts from IBM. These two datasets are suitably fused and integrated to generate localized meteorological alerts and "watch-over" episodic events in 2018. C-SAG successfully monitored and alerted farmers of the Titli and Phethai cyclones on October 10, 2018, and December 17, 2018, respectively, which helped farmers prepare in advance for the oncoming cyclones.

In addition, C-SAG could rapidly generate potential damage assessments at the village level and provide this information in just 48 hours. Further, C-SAG has developed automated methods to analyze wind rose, peak meteorological events spatially and seasonally, calibrated hot spots, micro-level patterns in meteorological data, and many other valuable techniques. Moreover, C-SAG is studying and developing a machine learning algorithm to fuse multi-spectral images and meteorological alerts at the village level, thereby showing spatial analytics' capability for data usage.

Peak meteorological (Figure 15.1) events are extreme weather phenomena that occur with high intensity within a given region. These events include, but are not limited to, heavy rainfall, thunderstorms, heatwaves, blizzards, and strong winds. Peak meteorological events are significant because they can cause severe damage to infrastructure, disrupt transportation and communication networks, and pose risks to human safety. Monitoring and analyzing peak meteorological events help improve forecasting accuracy, enhance disaster preparedness, and mitigate the impacts of extreme weather on communities and ecosystems.

A wind rose (Figure 15.2) is a graphical tool used to visualize the frequency and direction of wind at a specific location over a certain period. Typically, the wind rose is divided into sectors representing different compass directions (e.g., north, south, east, west), and the length of each segment indicates the frequency of wind blowing from that direction. The wind rose provides valuable information for various applications such as urban planning, agriculture, and renewable energy projects.

C-SAG has clearly embedded important weather inputs in agriculture advisories. As mentioned, these inputs are being provided using advanced GIS techniques to 53,000 farmers in the 2018 season in southern Odisha across 532 villages, and it is proposed to increase it to 133K farmers in the 2019 and 2020 seasons. C-SAG's primary focus is to generate more reliable ensembled weather

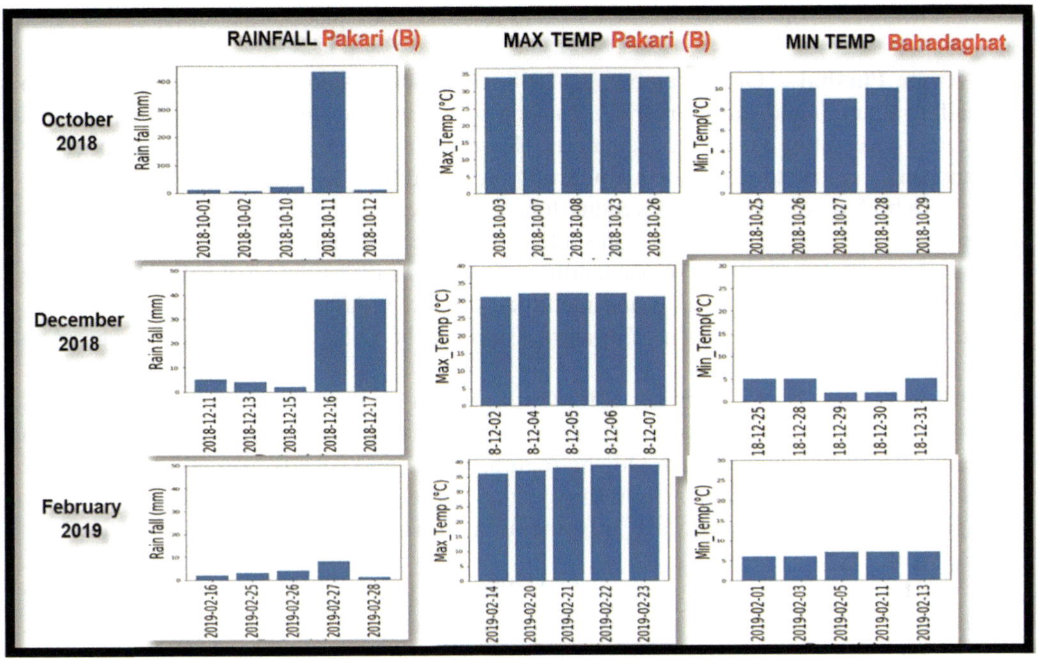

FIGURE 15.1 Peak events (identifying villages).

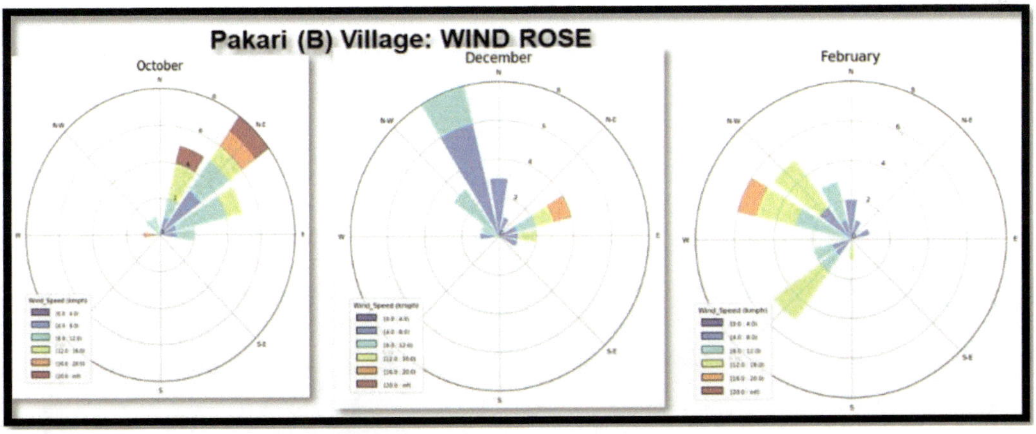

FIGURE 15.2 Wind rose for Pakari village.

products for its Agri-GIS and, therefore, generate beneficial advisories to plots/farmers from the datasets received from the Indian Meteorological Department and IBM.

In this regard, C-SAG has been working with IMD and IBM to undertake a unique micro-climatic analysis in southern Odisha by looking at gridded data at 3 kilometers for 2018, 2017, and 2016.

Therefore, the IBM team has been working with C-SAG in testing and deploying this enhanced weather forecast. C-SAG is now using an ensemble approach in weather advisory, and it goes without saying that having the datasets from IMD as well as weather data from IBM leads to better leveraging and therefore generating more trustworthy and accurate insights as well as forecasts.

The following are examples as to how C-SAG's forecasts took place.

1. **Cyclone Phethai** – On December 17, 2018, this tricky cyclone hit the Indian coast in Andhra Pradesh and started rushing, causing severe weather imbalance and damage. All weather forecasting services raised alerts. However, in between, the cyclone unexpectedly changed its course towards north Andhra Pradesh and Odisha (Figure 15.3).

   With high-quality enhanced weather forecasts provided by IMD and IBM, C-SAG successfully analyzed the trajectory and intensity of the Phethai cyclone and generated reliable alerts a day before the cyclone landing – December 17, 2018. C-SAG also successfully assessed the impact of Phethai and identified the likely damage to crops like red gram, paddy, groundnut, and cotton. The alert issues on the C-SAG portal can be viewed in Figure 15.4.
2. **Cyclone Titli** – On October 10, 2018, Cyclone Titli hit the Indian coast in Odisha and soon intensified into a severe cyclonic storm. This cyclone mainly affected Odisha state – primarily the southern Odisha region. With wind speeds over 140 km per hour and tides more than 1 meter high, the cyclone caused a high-impact landfall with very high to extremely high rainfall. In some areas, it ranged up to 225 mm of rain daily (Figure 15.5).

   The impact of the cyclone was so high that the Odisha state government ordered the closure of schools, colleges, and anganwadi centers (rural childcare centers) in Rayagada, Gajapati, Ganjam, Puri, and Jagtsinghpur districts.

   The weather data available by the Weather Company and IMD was of great value to C-SAG. It helped them monitor the cyclone's trajectory and generate village-wise alerts of severe weather advisory conditions to farmers. In addition, hyper-local weather alerts

FIGURE 15.3 Severe cyclonic storm Phethai over Bay of Bengal, located near 14.2N 82.2E at 161800z (162330IST).

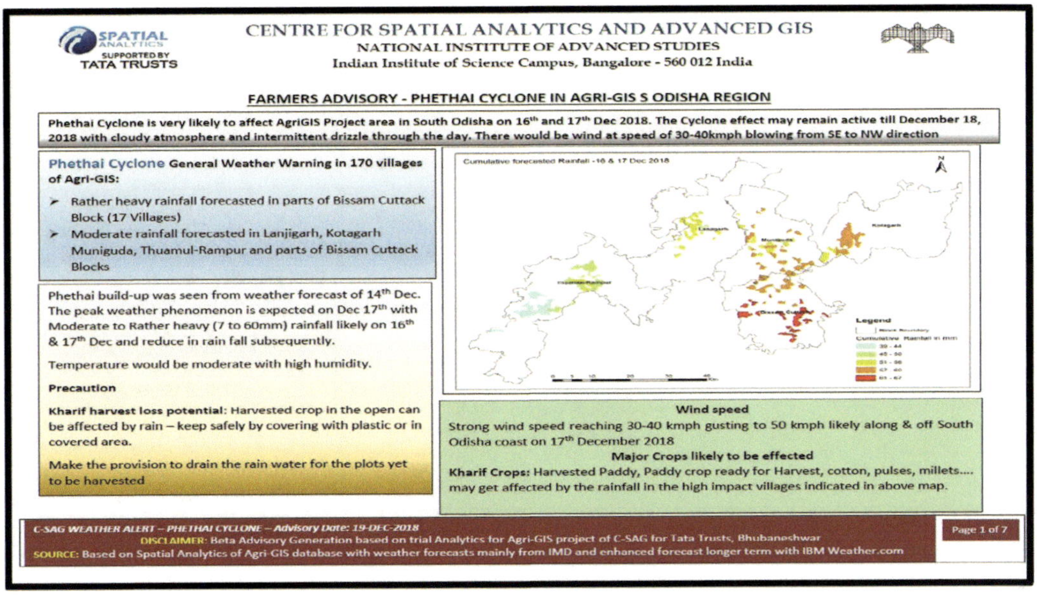

FIGURE 15.4 Phethai alert from C-SAG (showing villages impacted).

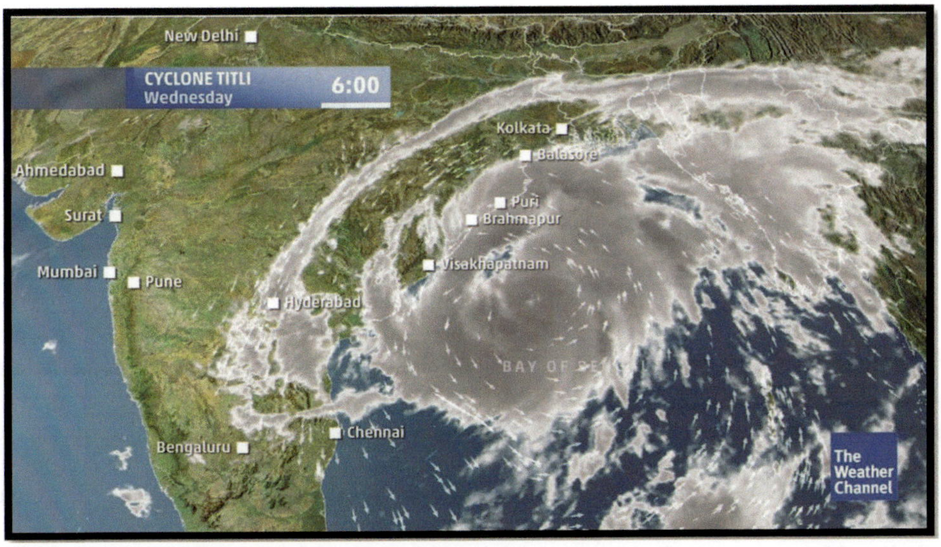

FIGURE 15.5 Cyclone Titli.

generated by C-SAG enabled an immediate local assessment of potential damage to villages, especially crops and animal lives. The in-time advisory just within one day of the cyclone was generated by C-SAG and was very useful for local administration (Figure 15.6).

15.3 IMPACT

C-SAG regularly accesses weather data from the Weather Company, an IBM business, and plans to conduct in-depth research using the data. They have started curating select weather data daily and

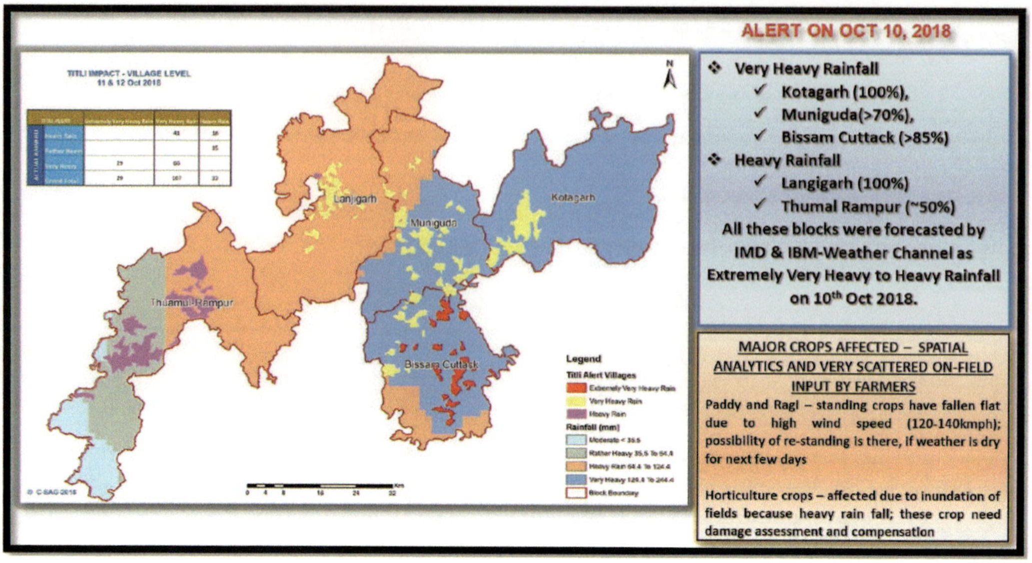

FIGURE 15.6 Titli alert from C-SAG (showing villages impacted).

have also begun to develop customized analytics, which would ultimately benefit farmers, at least at the village level, with many weather analytics products generated by C-SAG's team.

Early cyclone warnings are instrumental in facilitating the timely evacuation of people to safe places, thereby minimizing the risk of loss of life and property. Here's how early cyclone warnings help in moving people to safe places:

- **Advance Notice**: Early warnings provide communities with advance notice of an impending cyclone, allowing authorities to initiate evacuation procedures well before the cyclone makes landfall. This gives people adequate time to prepare and evacuate to safer locations.
- **Evacuation Planning**: Authorities use cyclone forecasts to develop evacuation plans and identify vulnerable areas that are at risk of storm surges, flooding, or landslides. This planning helps prioritize evacuation efforts and ensures that resources are allocated efficiently to evacuate those most at risk.
- **Public Awareness**: Early warnings are accompanied by public awareness campaigns and communication efforts to educate residents about the potential impacts of the cyclone and the importance of evacuating to safe shelters. This information empowers people to make informed decisions and proactively protect themselves and their families.
- **Shelter Management**: Cyclone warnings enable authorities to activate cyclone shelters and temporary evacuation centers before the storm. These shelters are equipped with essential supplies, emergency services, and trained staff to accommodate evacuees safely until the cyclone threat subsides.
- **Transportation Arrangements**: Local authorities arrange transportation, including buses, boats, and other vehicles, to facilitate the evacuation of residents from high-risk areas to designated shelters. Special arrangements are made for vulnerable groups such as the elderly, children, and people with disabilities to ensure their safe evacuation.
- **Coordination and Monitoring**: Government agencies, emergency responders, and community volunteers coordinate evacuation efforts and monitor evacuation routes to ensure smooth traffic flow and evacuees' safety. Real-time communication and updates are provided to keep residents informed throughout the evacuation process.

Overall, early cyclone warnings are crucial in saving lives by enabling timely evacuation and ensuring that people are moved to safe places before the onset of extreme weather conditions associated with cyclones.

Further, timely and accurate cyclone warnings were crucial for people residing in those areas, particularly the farmers engaged in the agriculture sector, in terms of the following reasons.

- **Crop Protection**: Cyclones can bring heavy rainfall, strong winds, and storm surges, damaging crops and agricultural infrastructure. Timely warnings allow farmers to take preventive measures such as harvesting crops early, securing farm equipment and structures, and reinforcing crop protection measures to minimize damage.
- **Risk Mitigation**: Farmers can use cyclone forecasts to assess the risk to their crops and livestock and take appropriate actions to minimize losses. This may include moving livestock to safer areas, stocking up on essential supplies, and reinforcing agricultural structures to withstand strong winds and heavy rainfall.
- **Resource Management**: Cyclone warnings help farmers plan their agricultural activities more effectively. For example, they can adjust their planting and harvesting schedules based on forecasted weather conditions to optimize resource allocation and reduce the risk of crop losses.
- **Livelihood Protection**: Agriculture is the primary source of livelihood for millions of farmers in India. Timely cyclone warnings enable farmers to protect their crops and assets, safeguarding their income and livelihoods from the adverse impacts of extreme weather events.
- **Government Support**: Government agencies and disaster management authorities use cyclone warnings to implement preparedness and response measures, including evacuations, relief distribution, and post-disaster recovery efforts. This coordinated response helps minimize the socio-economic impacts of cyclones on rural communities and supports farmers in rebuilding their livelihoods after a disaster.

Overall, accurate cyclone warnings play a critical role in enhancing the resilience of agriculture and farmers in India, helping them adapt to and mitigate the impacts of climate-related risks on their livelihoods.

BIBLIOGRAPHY

Gangopadhyay, Prasun K., Arun Khatri-Chhetri, Paresh B. Shirsath, and Pramod K. Aggarwal. "Spatial targeting of ICT-based weather and agro-advisory services for climate risk management in agriculture." *Climatic Change* 154, no. 1 (2019): 241–256.

Singh, K. K., A. K. Baxla, Priyanka Singh, and P. K. Singh. "Weather based information on risk management in agriculture." In *Climate change and agriculture in India: Impact and adaptation*, pp. 207–216. Cham: Springer International Publishing (2018).

Sivakumar, Mannava V. K., and Raymond P. Motha. "Managing weather and climate risks in agriculture summary and recommendations." In *Managing weather and climate risks in agriculture*, pp. 477–491. Berlin, Heidelberg: Springer, 2007.

16 Long-Term Rainfall Forecasting Using Data Science

Jayavaradhan Sambedu

ABSTRACT

Rainfall forecasting can be categorized into two types: short term and long term. Short-term forecasting has higher accuracy due to the low probability of sudden changes in the system. Apart from the agricultural sector, long-term rainfall forecasting also has advantages over other sectors. The Indian Meteorological Department (IMD) uses empirical and dynamic approaches for building long-term forecasting models.

Machine learning approaches are data driven, but until now, they haven't been incorporated to build long-term rainfall forecasting models by the IMD. In this chapter, Curl Analytics uses machine learning and time series techniques to forecast rainfall one and three months ahead of the current time, providing good accuracy.

16.1 RAINFALL FORECASTING IN INDIA

Rainfall forecasting has been an essential field of study for ages. Over time, the approaches evolved with knowledge of the science behind atmospheric dynamics. For India, rainfall forecasting is of immense importance given that we are still primarily an agrarian society, and much of our agriculture is rain fed.

In India, both short-term and long-term rainfall forecasts are crucial due to the country's diverse climate and its heavy reliance on agriculture. A brief overview of both methods is given in the following.

Short-term rainfall forecasts in India typically cover a period of a few hours to a few days. These forecasts are essential for immediate decision-making in various sectors, including agriculture, disaster management, and urban planning. They rely on weather radar data, satellite imagery, numerical weather prediction models, and ground-based observations. Short-term forecasts help predict localized heavy rainfall, thunderstorms, and flash floods, enabling authorities to issue timely warnings, evacuate vulnerable areas, and implement flood control measures.

Long-term rainfall forecasts in India extend over weeks, months, or seasons. These forecasts are vital for agricultural planning, water resource management, and climate adaptation strategies. They are based on climate models, historical weather data, oceanic and atmospheric conditions (such as the Indian Ocean Dipole and the El Niño-Southern Oscillation), and teleconnections. Long-term forecasts provide insights into seasonal precipitation patterns, drought risks, and the likelihood of above-normal or below-normal rainfall conditions. They help farmers decide crop selection, planting schedules, irrigation planning, and water conservation measures. Additionally, long-term forecasts assist policymakers in formulating drought mitigation strategies, allocating water resources, and implementing climate-resilient infrastructure projects.

Apart from the agricultural sector, long-term rainfall forecasting has advantages in other sectors as well. The financial sector heavily relies on rainfall forecasting for better macro-level planning, power generation through dams can be controlled if the forecast is available, water resources for drinking can be managed during drought years, various disaster management boards can prepare for flood

DOI: 10.1201/9781003487531-20

situations if above-normal rainfall forecast information is available to the organization (Uttarakhand in 2013 and in Kerala in 2018 where both loss of property and life were at very high levels), rainwater catchment systems are more vulnerable and less reliable when located where rainfall is highly variable in space and time, forecasting well before the season will help to strategize these systems, and merchant ship routes can be planned to avoid bad weather in the sea to avoid mishaps.

Short- and long-term rainfall forecasts are essential for managing weather-related risks, optimizing agricultural practices, and enhancing resilience to climate variability and change in India. They enable stakeholders to make informed decisions and adapt to evolving environmental conditions, ultimately contributing to sustainable development and food security.

16.2 DATA SCIENCE FOR RAINFALL FORECASTING

With data science and the Fourth Industrial Revolution, almost all industries will follow data-driven approaches to find the solution to real-life problems. Data science has gained mileage since the beginning of this decade. The world produces more and more data, and because of the development in data storage technologies, all the produced data can be stored cheaply. Consequently, many algorithms have been developed to learn patterns from the data to help science, technology, and businesses grow and make better decisions based on previous events or patterns. Data science has various branches, such as machine learning, statistics, deep learning, artificial intelligence, and time series, which can solve multiple business-related problems such as creating recommendation systems for e-commerce websites, building automated self-driving cars, suggesting optimal road routes to save time, helping businesses forecast demand for a product, helping intelligence agencies predict the next vulnerable zone for attack, helping banks predict fraud, and helping insurance companies strategize policies. Data science solves all these problems.

Rainfall forecasting, just like the challenges mentioned previously, is primarily solved by data science techniques. Curl Analytics demonstrated that long-range rainfall forecasting problems were solved with reasonable accuracy for India, China, Brazil, and Nigeria. We solved long-range rainfall forecasting at a country level, and now we are also focused on building models for district levels. To do so, we were selected among 2000 contenders by Karnataka Biotechnology and Information Technology Services (KBITS) with a grant to carry forward our work.

The Indian Meteorological Department (IMD), a government body that takes care of all the weather-related matters of India, is very successful in building short-term rainfall forecasting models. However, IMD could not forecast some draught years for long-term rainfall forecasting models. IMD uses empirical approaches to forecast rainfall and monsoons for long-term forecasting. Often, such parametric techniques are designed in such a way that they are best suited to empirical approaches to estimate short-term forecasts and not necessarily long-term ones.

16.2.1 Need for Long-Term Rainfall Forecast

As highlighted, rainfall forecasting can be categorized into two types: short term and long term. Short-term forecasting has higher accuracy due to the low probability of sudden changes in the system. For example, the probability of rain in the next hour is very low if it is not raining right now. However, forecasting whether it will rain in the next hour needs many dynamic parameters from land weather stations, satellites, and balloons. Atmospheric dynamics are complex, and a heavy numerical model is required, which uses differential equations to run on supercomputers. In contrast, long-term forecasts are less accurate and dynamic than short-term forecasts. They require less data, different variables from all over the world, and, at the same time, far less complicated computation. Finally, they require empirical (based on data) rather than numerical approaches.

IMD has many dynamic or short-term forecasting models with high accuracy that can be used daily. However, long-term forecasts are needed because long-term forecasting models failed to forecast deficient all-India southwest monsoon in 2002, 2004, 2009, and 2014. Simply put, the models employed by IMD were unable to predict drought. The drought was evident because the rainfall was lower than the –10% long-period average (LPA) for these years. Therefore, long-term forecasting models should be updated and deployed using machine-learning techniques.

India's southwest monsoons (usually from June to September) are among the most critical agricultural seasons, as India receives about 70–90% of its total rainfall. Farmers in India usually grow kharif crops during this period, accounting for a significant share of the gross domestic product (GDP). Agriculture contributes 15–18% of the total GDP currently, which has dropped from an almost 50% share in the GDP in the 1950s, while the service sector's contribution to the GDP has drastically increased in the last couple of decades. Agricultural output for India is highly correlated with the amount of rainfall received during the southwest monsoon period. This means that the

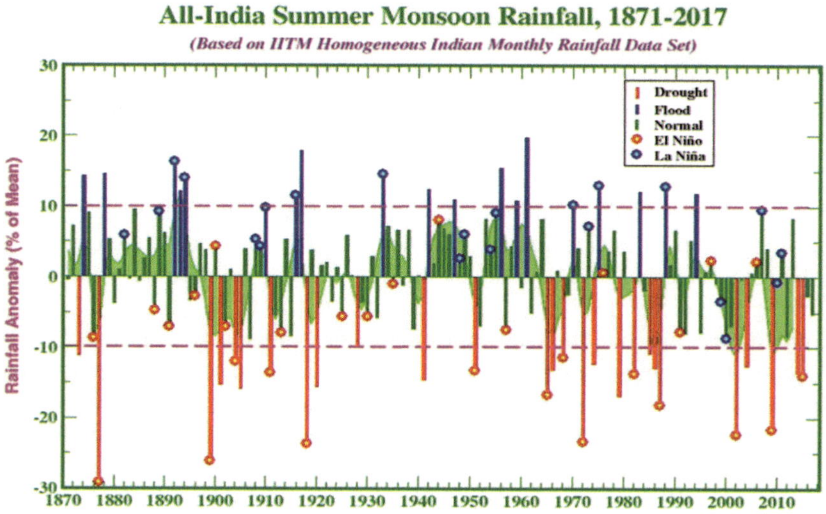

FIGURE 16.1 Variation in monsoon rainfall over 100 years in India.

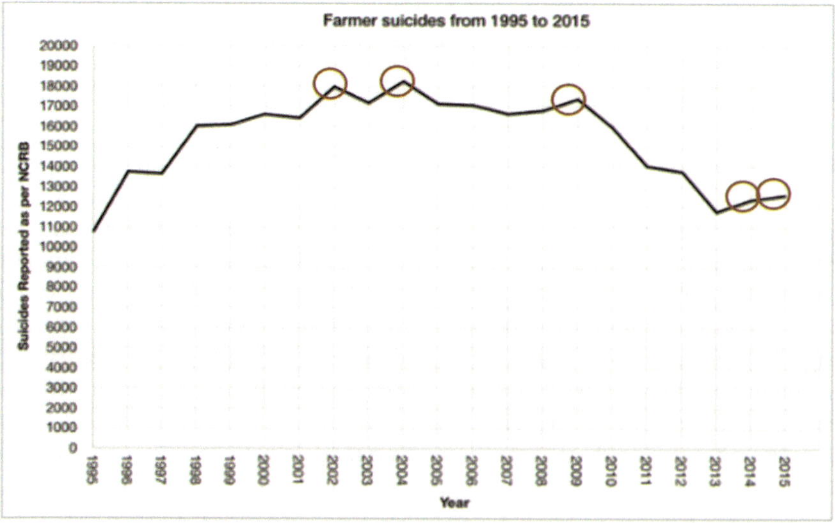

FIGURE 16.2 Mapping farmer suicides with drought years in India.

Indian agrarian economy heavily depends upon the all-India southwest monsoon rainfall. Although, in technical terms, the southwest monsoon is stable and regular, with ±30% of the long period average, only ±10% deviation from it causes a considerable GDP loss of 2–5% during the year (Figure 16.1). Further, a poor monsoon results in substantial financial debts for farmers. The number of farmer deaths during drought is correlated with all-India southwest monsoon rainfall during those years. Figure 16.2 shows a high correlation between farmer suicides in the drought years, that is, 2002, 2004, 2009, 2014, and 2015.

If the long-term rainfall forecasting models are strong enough to forecast drought periods well before drought events, such losses can be controlled to a greater extent.

Currently, IMD uses two approaches for building long-term forecasting models, empirical and dynamic. IMD itself believes that empirical approaches work much better than dynamical approaches. Thus, machine learning approaches are data driven, but until now, they haven't been incorporated to build long-term rainfall forecasting models by IMD. In our project, we build machine learning and deep learning models for forecasting long-term rainfall at the local level.

16.2.2 Rainfall Forecasting at the Indian Meteorological Department

In general, IMD uses two techniques to forecast long-range rainfall. The first method is a data-driven approach that uses the historical relationship between All-India SW monsoon rainfall and its predictors from global patterns of atmosphere and ocean parameters. The second method is the dynamical model approach, based on the global Climate Forecasting System Version 2 (CFSV2). The CFSV2 is a fully coupled general circulation model (CGCM) implemented by the Indian Institute of Tropical Meteorology (IITM) under the Monsoon Mission project launched by the Ministry of Earth Sciences (MoES), Government of India. The operational model for long-term forecast is the statistical or empirical (data-driven) model in IMD because such models have performed better than dynamic models in terms of long-term rainfall forecast. Dynamic models have not shown promising results to accurately simulate the salient features of the mean monsoon and its interannual variability (stated in one of the papers published by the National Climate Centre, Indian Meteorological Department). In short, empirical models perform much better for long-term rainfall forecasts than numeric models. Thus, machine learning techniques are all data driven, and these techniques try to capture the pattern in the data precisely because the algorithms are designed in such a way.

The following are the two critical features of long-term rainfall forecasting at IMD:

- IMD publishes long-term forecasts at four broad levels (northwest India, northeast India, central India, and south peninsula). In addition, it issues forecasts for monthly rainfall (for July and August) and for the second half of the season (August and September) over the country using models based on the latest empirical techniques.
- In addition to the quantitative forecast, the long-term forecasting system by IMD is also used to generate a five-category probabilistic season rainfall forecast based on the forecast error distribution of the forecasting system. The five categories defined based on the observed all-India SW monsoon rainfall data for the period 1901–2005 are deficient (<90% of LPA), below normal (90–96% of LPA), normal (96–104% of LPA), above normal (104–110% of LPA), and excess (>110% of LPA).

The techniques used by IMD are based on algorithms developed in data science and are advanced enough to capture the pattern for long-term rainfall forecasts. However, the models and techniques in data science have higher abilities to understand the patterns in the data and tell which of the five rainfall categories will apply for the region. IMD also tried to use artificial neural networks to find rainfall patterns in the data. We improve upon the artificial neural network models by building complex neural network architectures such as recurrent neural networks and memory-based recurrent neural networks to solve many time-series problems (rainfall is a time-series problem). We will explain the approach used for rainfall forecasting.

16.2.3 Rainfall Forecasting Solution

We took the real-world challenge of solving the problem of long-term forecasts. We developed a model using advanced machine learning and time series techniques to forecast one and three months ahead of the current time. Many reputed meteorological organizations state that long-term rainfall forecasting depends more on empirical than numerical solutions. We tried to use the best data-driven models to forecast rainfall. We did extensive back tests of around 115 years of data. The model showed good accuracy for India, primarily because the variables chosen were based on understanding the atmospheric dynamics of the Indian subcontinent. For example, variables such as pressure and temperature worldwide were selected to build the model. The locations chosen for the modeling are shown in Figure 16.3 and the data points used in Figure 16.4.

In the initial stage, the model was built to forecast rainfall for Brazil because the US market heavily depends on Brazil's agricultural produce. Hence, readily available validation exercises can be used to test our model. For example, Brazil exports a significant chunk of coffee production worldwide, and many markets depend upon the amount of coffee imported from Brazil. We used rainfall as one of the key indicators of how much coffee Brazil will export to the US. Moreover, this information was used to understand the behavior of other food companies and big cafés worldwide. The success of rainfall forecasting for Brazil motivated us to build a model for Indian rainfall, too. Using variables precisely according to the dynamics of the atmosphere according to the Indian region, the model performed better than the Brazil model. The results showed high accuracy for India, China, Brazil, and Nigeria. Thus, we could determine if the rainfall in the next month and next three months would be *below normal*, *normal*, or *above normal* at the country level. The

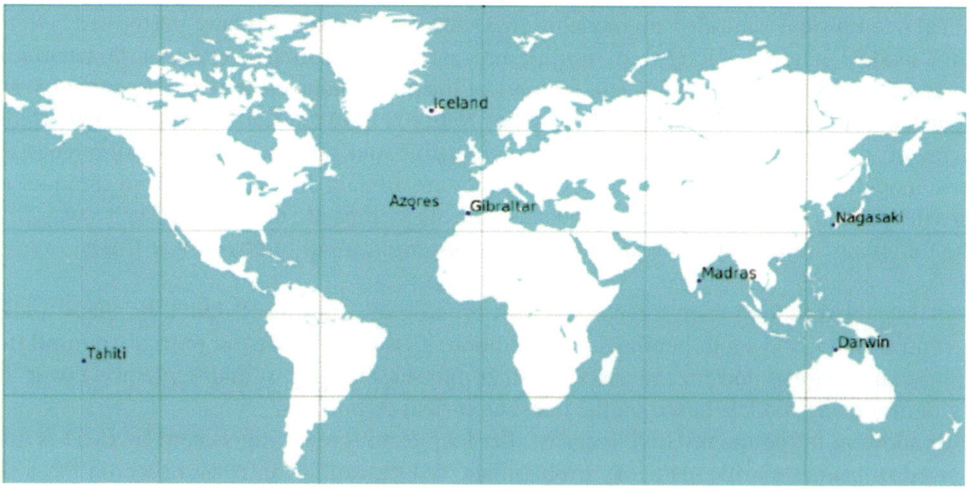

FIGURE 16.3 Locations where pressure data was taken for the models.

Input Variables	
Sea Level Pressures	**Ocean Pressure Indices**
Madras	Southern Oscillator Index (SOI)
Nagasaki	North Atlantic Oscillator Index (NAO)
Darwin	Pacific Decadal Oscillator (PDO)
Tahiti	Arctic Oscillator (AO)
Gibraltar	North Pacific Index (NPI)
Iceland	**Temperature of the region**
Azores	**Past rainfall of the region**
	Trend and seasonality inputs

FIGURE 16.4 Input variables used for the model.

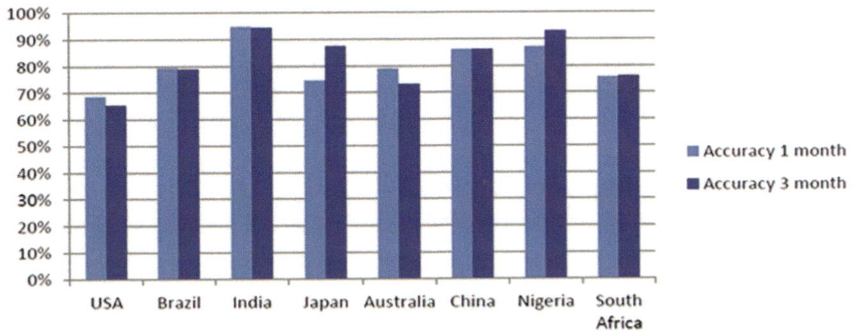

FIGURE 16.5 Showing the model's accuracy across different countries.

accuracy of our model for a few selected countries is shown in Figure 16.5. Accuracy for India is more than 90% for one month and three months ahead of forecast.

16.3 NEXT STEPS OF THE PROJECT

After success with long-term rainfall forecasting, Curl was selected among 2000 contenders for the KBITS grant to improvise the long-term rainfall forecasting model. Karnataka State Natural Disaster Monitoring Center (KSNDMC) will be helping Curl Analytics with the required data. We are in talks with the Indian Meteorological Department to get the necessary data.

Our next goal is to build a model with a hierarchical approach. We have already built a model at the country level. We would now be configuring the model to higher granularity like states and districts to forecast three months before rainfall. This model will help farmers and businessmen. Farmers can plan their crops accordingly by micro-level forecasts and not only macro-level forecasts. We also plan to use artificial intelligence models to understand complex atmospheric dynamics, make the model robust, and give accurate predictions. The following steps shows the hierarchical plan for forecasting.

Our next step will also include the automation process, where we will run the end-to-end process from data collection to information sharing with the end user. Our automation process will have the following steps:

- Internet of Things (IoT): The IoT will extract data using weather sensors installed at various places. Communication at the sensor level will automatically pass live data into the database present in the cloud.
- Artificial intelligence–based models: We will be building models based on machine learning, deep learning, and artificial intelligence by extracting the data present in the cloud database and pushing back the results to the cloud database itself.
- Cloud-based solutions: Forecasts for the end user will be available through web-based mobile phone and computer applications.

BIBLIOGRAPHY

Dash, Yajnaseni, Saroj K. Mishra, and Bijaya K. Panigrahi. "Rainfall prediction for the Kerala state of India using artificial intelligence approaches." *Computers & Electrical Engineering* 70 (2018): 66–73.

Hussein, Eslam A., Mehrdad Ghaziasgar, Christopher Thron, Mattia Vaccari, and Yahlieel Jafta. "Rainfall prediction using machine learning models: Literature survey." *Artificial Intelligence for Data Science in Theory and Practice* (2022): 75–108.

Kumar, Vijay, Sharad K. Jain, and Yatveer Singh. "Analysis of long-term rainfall trends in India." *Hydrological Sciences Journal–Journal des Sciences Hydrologiques* 55, no. 4 (2010): 484–496.

Manandhar, Shilpa, Soumyabrata Dev, Yee Hui Lee, Yu Song Meng, and Stefan Winkler. "A data-driven approach for accurate rainfall prediction." *IEEE Transactions on Geoscience and Remote Sensing* 57, no. 11 (2019): 9323–9331.

17 Sundarbans Mangrove Tigers

How They Came to Be Camera-Trapped and Counted

Debmalya Roy Chowdhury and Sampurna Sarkar

ABSTRACT

Since 2006, tiger estimation exercises using camera traps have been conducted every four years in India under the supervision of the National Tiger Conservation Authority (NTCA). Tiger estimation exercises using camera traps involve deploying remote cameras equipped with motion sensors in tiger habitats to capture images of tigers and other wildlife. These camera traps are strategically placed across known tiger habitats and movement corridors to maximize the chances of detecting tigers and obtaining accurate population estimates. The collected images are then carefully analyzed by trained experts to identify individual tigers based on their unique stripe patterns. This allows researchers to estimate tiger populations and monitor population trends over time. Camera trap surveys provide valuable data on tiger distribution, abundance, and demographics, which are essential for conservation planning, habitat management, and monitoring the effectiveness of conservation efforts. These surveys also help identify priority areas for conservation interventions and guide conservation strategies to protect and conserve tiger populations in their natural habitats. This chapter highlights how camera traps were used to capture data and further used to estimate the number of tigers using statistical sampling methods in the Sundarbans Tiger Reserve in 2012.

ABOUT THE SUNDARBANS

The Sundarbans is a vast mangrove forest in the Bay of Bengal delta region, spread across India and Bangladesh. It is one of the world's most extensive tidal halophytic mangrove forests and is renowned for its rich biodiversity, including the iconic Bengal tiger. The region is characterized by its unique ecosystem, where saltwater from the sea mixes with freshwater from rivers, creating a complex network of tidal waterways, mudflats, and small islands. The Sundarbans is also home to a diverse array of flora and fauna, including numerous species of birds, reptiles, and marine life. Additionally, the area supports a large population of local communities who depend on its resources for their livelihoods, primarily through fishing and honey collection. However, the Sundarbans face numerous environmental challenges, including habitat loss, climate change, and human-wildlife conflict, which threaten its ecological integrity and the well-being of its inhabitants. UNESCO designates this eco-sensitive estuarine delta as a World Heritage Site. Lying between latitude 21° 27′ 30″ and 22° 30′ 00″ North and longitude 89° 02′ 00″ and 90° 00′ 00″ East and with a total area of approximately 10,200 square kilometers, 60% of it lies in Bangladesh and the rest in India. The land area, including exposed sandbars, occupies 414,259 ha (70%), with water bodies covering 187,413 ha (30%). In the Indian Sundarbans ecoregion, there are 102 deltaic islands, of which 54 are variably inhabited, totaling about 4.5 million (2011, Census). A map of the Sundarbans is shown in Figure 17.1.

DOI: 10.1201/9781003487531-21

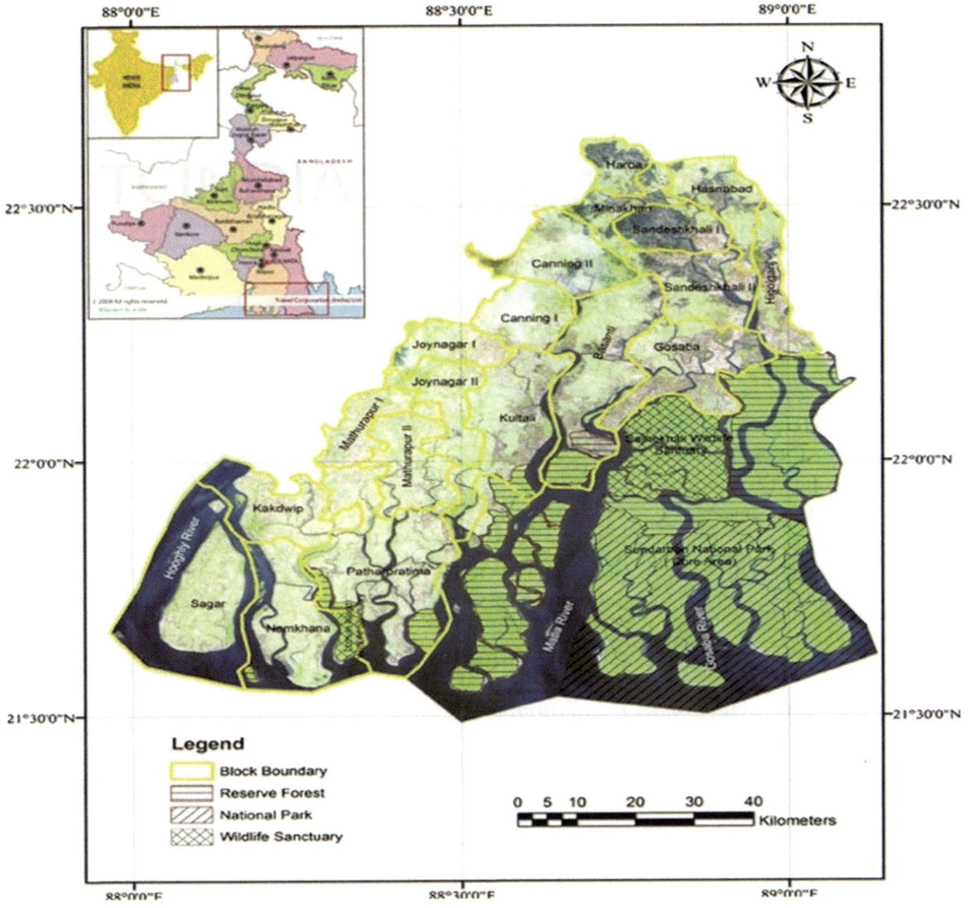

FIGURE 17.1 Location map of Indian Sundarbans (source: Dasgupta et al.).

17.1 UNDERSTANDING THE SUNDARBANS FOREST: HABITAT, FLORA AND FAUNA

The Sundarbans Tiger Reserve (STR) is situated in the coastal districts of West Bengal, that is, South 24-Parganas and part of North 24-Parganas. The Sundarbans can be described as a maze of estuaries, river channels, and creeks encompassing a number (102) of islands of various shapes and sizes. It is the only mangrove forest worldwide (besides Bangladesh) to harbor a significant tiger population. Apart from the tiger, the Sundarbans Tiger Reserve also has a rich array of biodiversity values, both floral and faunal. The Sundarbans Tiger Reserve is bound on the east by the international boundary with Bangladesh formed by the rivers Harinbhanga, Raimangal, and Kalindi. On the south lies the Bay of Bengal. The western border is formed by the river Matla, a common boundary with the territorial Forest Division of South 24-Parganas. Towards the northwest, the area is bound by rivers Bidya and Gomdi.

The core area extends up to 1699.62 sq. km. and is devoid of any human settlement, with no human activities (fishing, crabbing, etc.) allowed inside, while the buffer extends up to 885.27 sq. km. Both tourism and fishing are permitted in a controlled manner in the buffer zone.

The vegetation is predominantly mangroves. Besides the forest, there are extensive areas of brackish water and freshwater marshes, intertidal mudflats, sandflats, sand dunes with typical dune vegetation, open grassland on sandy soils, and raised areas supporting a variety of terrestrial shrubs and trees. This unique mangrove ecosystem, with its numerous ecological niches, is home to over 1586 faunal species; among them, 15 mammals, 8 species of birds, and 17 species of reptiles are included in Schedule I and II of the Wildlife (Protection) Act, 1972.

CITES stands for the Convention on International Trade in Endangered Species of Wild Fauna and Flora. It's an international agreement between governments to ensure that international trade in wild animals and plants doesn't threaten the species' survival. CITES lists species under three appendices, with Appendix I containing the most endangered species threatened with extinction. As of the last update in September 2021, the 14 species listed in Appendix I of CITES are:

- Asian elephant (*Elephas maximus*)
- Javan rhinoceros (*Rhinoceros sondaicus*)
- Sumatran rhinoceros (*Dicerorhinus sumatrensis*)
- Tiger (*Panthera tigris*)
- Snow leopard (*Panthera uncia*)
- Cheetah (*Acinonyx jubatus*)
- Lion (*Panthera leo*)
- Polar bear (*Ursus maritimus*)
- Gorilla (*Gorilla* spp.)
- Chimpanzee (*Pan troglodytes*)
- Bonobo (*Pan paniscus*)
- Fin whale (*Balaenoptera physalus*)
- Blue whale (*Balaenoptera musculus*)
- Humpback whale (*Megaptera novaeangliae*)

17.2 SWAMP TIGERS

There is widespread interest among scientists and conservationists regarding the origin of tigers in the Sundarbans. It is usually believed that the leaner frame and lower body mass [Sundarbans tigers (females) weigh 75–80 kg, whereas mainland tigers (females) weigh 100–160 kg] possibly imply that Sundarbans tigers are different from the tigers in the rest of India.

The Sundarbans ecosystem is governed by salinity and tides. Therefore, the life of terrestrial organisms here is not the same as their counterparts elsewhere. The mangrove terrain is subjected to tides twice a day. A combination of daily tidal water fluctuations induced by the lunar cycle and changes in water flow and level during seasons result in periodic inundation of land. For some areas, the inundation is twice daily; for some other places, fortnightly; and for different regions, only during the monsoon season.

The changing water level regulates biological diversity. Though mangrove habitats are amongst the most productive ecosystems, they can be challenging for terrestrial mammals, including tigers, due to the lack of fresh water and difficulties associated with movement across the swampy terrain with dense vegetation. Due to limited freshwater flow, vegetation in the Indian Sundarbans is stunted and has a low canopy. The presence of tigers makes the Sundarbans mangrove ecosystem unique.

Until about two decades ago, the estimation of tiger numbers in the mangrove habitat was based on tiger pugmarks, statistics of tiger attacks on humans, and residents' experiences. Finding pugmarks on a landmass, about 30% of which goes underwater twice daily, is no easy task. Further, the same pugmark may vary in dimension before and after tidal impact. The estimates based on the pugmark method in the Sundarbans were much debated.

In 2000, Dr Ullas Karanth, notable for pioneering the scientific use of camera traps in population density studies of large wild mammals in India, placed camera traps along human-made freshwater ponds in the Indian Sundarbans and estimated the presence of 100 tigers (Karanth and Nichols, 2000). Extrapolation for the entire mangrove habitat was a significant limitation of the study. He concluded that conventional camera trapping was unsuitable for estimating tiger numbers in the Sundarbans.

The conventional grid-based camera trapping method requires the placement of camera traps in every grid of the sampling area. However, for a complex mosaic of landmass and water, like the Sundarbans, conventional camera trapping method is unsuitable due to:

1. Inaccessibility of some grids,
2. Some grids within the sampling area overlap with water channels,
3. Absence of animal trails,
4. Low elevation within the sampling area where camera traps may be inundated during high tide, and
5. Long sampling period.

Since 2006, tiger estimation exercises using camera traps have been conducted every four years in India under the National Tiger Conservation Authority (NTCA) supervision. However, in the first such exercise, the Sundarbans were left out, and the first all-India tiger status report (Jhala et al., 2008) published by NTCA in 2008 did not report tiger numbers from the Sundarbans. The second all-India tiger status report, published in 2011, reported 70 tigers in the Indian Sundarbans, although no camera traps were installed during the exercise (Jhala et al., 2011). The estimate was based on radio-collaring of some tigers in the mangrove habitat. After the number was published, the principal chief conservator of forest–wildlife (PCCF-WL) and chief wildlife warden (CWLW) of West Bengal expressed reservations regarding the estimated number of tigers and the methodology. The CWLW advocated the development of a method appropriate for the Sundarbans.

The Lothian Wildlife Sanctuary (LWS), under the Bhagabatpur Range of 24 Parganas (South) Forest Division, was selected as the site for experimentation to develop an appropriate camera trapping methodology. LWS has the highest mangrove biodiversity in the Sundarbans and is bereft of tigers.

The experiment on LWS began in March of 2011 by placing ten pairs of camera traps that were thought to be best suited for camera trapping spotted deer, wild boar, jackal, and other small felines such as jungle cat, fishing cat, and leopard cat.

Within a week of the deployment of camera traps, a "supermoon" occurred. A supermoon is a full moon (or a new moon) that nearly coincides with perigee – the closest that the Moon comes to the Earth in its elliptic orbit – resulting in a slightly larger-than-usual apparent size of the Moon as viewed from Earth. The phenomenon occurs when the full moon's orbit is closest to the Earth's orbit. Due to the closer proximity of the Moon, the tides are higher than usual. All ten pairs of camera traps were inundated. Dr Karanth was vindicated, but the team was on a mission.

The team decided to redeploy camera traps and monitor the tides by camping nearby on Lothian Island. This was to remove camera traps immediately to safety if required. The Sundarbans revealed themselves to the camping team. Not all the island was inundated during high tide, and all wildlife moved to these high-elevation areas during high tide. This was the eureka moment. Now, the trick was to find high-elevation regions and the variation in tide throughout the year. If camera traps were installed at the highest locations and on the day of the highest high tide, the chances of inundation during the sampling period were negligible.

A tide table published by the Kolkata Port Trust was procured, and the data was analyzed to find the months and periods with the lowest high tide. A map of the island was overlaid with grids of one sq. km., and camera trap locations were selected at high-elevation areas.

In 2011, camera traps were installed on Lothian Island in two phases, first at 26 locations for 10 days and then at 10 other locations for 90 days. The height of the camera traps was determined, keeping in mind the height of the target species. Images of animals were recorded. The appropriate methodology was in sight, but pilferage was posing a challenge.

To protect the camera traps, a tamper-proof steel box was designed to be securely tied to trees with plastic-coated steel cables. The ends of the cable were secured using a customized steel lock that needed an Allen key of a specific size, which was not readily available in the Sundarbans.

The measures were successful in safeguarding the camera traps. More than 800 images of spotted deer, wild boar, small cats, jackals, and birds were captured, based on which a research paper by Das et al. (2017) was published, confirming the presence of about 14 fishing cat individuals on Lothian Island.

The camera trapping method was ready to be tested in the tiger-bearing mangrove forest. Authorization was granted for the Raidighi, Ramganga, and Matla ranges of the 24 Parganas (South) Forest Division.

Between January and March 2012, camera traps were placed at 41 locations in the Ramganga and Raidighi ranges in two phases spanning 64 days. An olfactory lure was used to attract tigers and prolong their presence in front of camera traps. The lure was developed by decomposing goat meat and eggs and sun-drying them for three to four days. The result was a semi-liquid mixture with the odor of animal carcasses. The olfactory lure was smeared on plants around the camera traps.

Camera traps were placed on the bunds of the freshwater ponds and in high-elevation areas. The presence of animal signs (hoofmarks, pugmarks, rake marks, scrape marks, scat, etc.) was considered while selecting the sites. Two cameras were placed facing each other, 40–45 cm above ground, 15–20 feet apart, to get a clear picture of the stripes of both flanks of the tiger. The intervening space was cleared of vegetation. Each camera trap station was located within a 4–sq. km. grid. The camera traps were checked periodically, and the olfactory lure was reapplied to maximize the recapture of tigers and other carnivore species.

Based on the unique stripe patterns of each tiger, the identity of 20 tiger individuals was established in the 24 Parganas (South) Forest Division. Until then, the division reported four to five tigers in its annual reports. The presence of these tigers led to the creation of the West Sundarbans Wildlife Sanctuary,[1] the largest in the Indian Sundarbans, spanning 556 square kilometers. The camera trapping method came to be adopted as the monitoring protocol for the Sundarbans by NTCA in 2012.[2]

The camera trapping method was adopted as the monitoring protocol for the Sundarbans by the National Tiger Conservation Authority due to its effectiveness in capturing images of elusive wildlife species, including tigers, in their natural habitat. Camera trapping allows for non-invasive monitoring by capturing images and videos of wildlife as they move through designated areas. This method provides valuable data on species presence, abundance, behavior, and distribution, aiding conservation efforts and informing management strategies. The protocol outlined by the NTCA provides guidelines for systematically deploying camera traps, data collection, image analysis, and reporting procedures to ensure standardized monitoring practices across the Sundarbans landscape. The camera trapping method continues to be in use, and the fifth all-India tiger report released in April 2023 confirmed the presence of 101 tigers in the Indian Sundarbans (Qureshi et al., 2023).

NOTES

1 https://www.wildbengal.com/west-sundarban.php. Accessed on 16 October 2023 at 13:41 hours.
2 https://ntca.gov.in/assets/uploads/guidelines/Monitoring_Sunderban.pdf. Accessed on 16 October 2023 at 13:43 hours.

REFERENCES

Das S. K., Saha R., Mukherjee S., Danda A. and Borah J. 2017. First estimates of fishing cat abundance and density in Lothian Wildlife Sanctuary, Sundarbans, India. *Cat News* 66, 25–27.

DasGupta, R. and Shaw, R. 2015. An indicator based approach to assess coastal communities' resilience against climate related disasters in Indian Sundarbans. *Journal of Coastal Conservation* 19(1), 85–101.

Jhala Y. V., Gopal R. and Qureshi Q. (eds). 2008. *Status of Tigers, Co-Predators and Prey in India.* National Tiger Conservation Authority and Wildlife Institute of India. TR08/001, p. 164.

Jhala Y. V., Qureshi Q., Gopal R. and Sinha P. R. (eds). 2011. *Status of the Tigers, Co-Predators, and Prey in India, 2010.* National Tiger Conservation Authority, Government of India, New Delhi, and Wildlife Institute of India, Dehradun. TR 2011/003, p. 302.

Karanth K. U. and Nichols J. D. 2000. *Ecological Status and Conservation of Tigers in India. Final Technical Report to the Division of International Conservation, US Fish and Wildlife Service, Washington DC and Wildlife Conservation Society, New York.* Centre for Wildlife Studies, Bangalore, India.

Qureshi Q., Jhala Y. V., Yadav S. P. and Mallick A. (eds). 2023. *Status of Tigers, Co-Predators and Prey in India, 2022.* National Tiger Conservation Authority, Government of India, New Delhi, and Wildlife Institute of India, Dehradun.

Section IV

Energy, Water and Sanitation

18 Integrated Energy Modelling for India

Simi Thambi and Anindya Bhattacharya

ABSTRACT

Energy modelling can contribute immensely to integrated energy planning and replicate real-world energy systems. As a result, national and even global models can be established. These models can be used to display alternative futures with different fuel supply options. They provide an integrated outlook to the different energy domains of coal, oil, gas, nuclear, and renewables.

Energy modelling in India is fragmented and has limitations. This chapter discusses the challenges and opportunities in using energy models effectively for policy making. Global best practices have been listed to demonstrate the status and importance of energy modelling across the world. Then we share insights gained from national and state-level energy modelling and inter-model comparison exercises to bring out their scope and some areas of opportunities. NITI Aayog has developed an in-house energy optimisation model. A scenario assessment between the baseline and Nationally Determined Contribution (NDC; presented before UNFCCC) scenarios shows that by achieving its targets for increasing renewable capacity per the NDC commitments, India can increase the share of renewables in power generation to 22% by 2032.

18.1 INTRODUCTION

Energy modelling contributes immensely to integrated energy planning and policy analysis. Energy models replicate real-world energy systems to create sub-national, national, regional, or global models. Once these models are developed, they can be used to create scenarios to display alternative futures with different technologies and fuel supply options. The models provide an integrated outlook to the energy domain, which straddles different sectors – coal, oil, gas, nuclear, and renewables.

The use of energy modelling for policy making in India is not new, but due to various reasons, these activities are yet to be fully integrated into institutional setup. This chapter discusses the challenges and opportunities in using energy models effectively for policy making. In the first part, global best practices are listed to accentuate the importance of energy modelling in other countries. In the second and third part, we share some insights gained from national- and state-level energy modelling and inter-model comparison exercises to bring out their scope and some areas of opportunities.

18.1.1 Global Best Practices in Using Energy Models to Inform Policy

Across the world, there are several countries that have government organisations that, by their legal mandate, are required to undertake modelling for the entire energy sector. United States has the Energy Information Administration (EIA), which is an independent statistical agency of the Department of Energy; the OECD countries have the International Energy Agency (IEA); China has the Energy Research Institute, which comes under the National Development and Reform Commission (NDRC); the European Commission has the Joint Research Centre (JRC). Through these organisations, modelling inputs regularly feed into policy outputs from time to time.

DOI: 10.1201/9781003487531-23

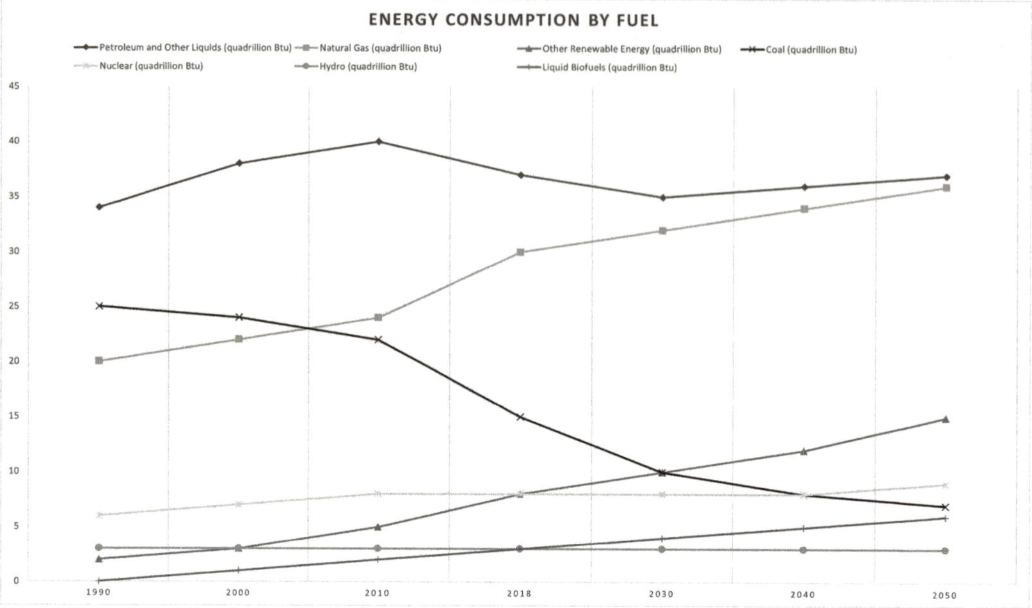

FIGURE 18.1 Energy consumption by fuel (data: Annual Energy Outlook: United States).

The EIA of the United States publishes three types of outlook: Annual Energy Outlook, Short Term Energy Outlook, and International Energy Outlook.

The Annual Energy Outlook (AEO) provides long-term energy projections for the United States every year. The projections are made until the year 2050. Projections in the Annual Energy Outlook are projections of what may happen given certain assumptions and methodologies. The AEO is developed using the National Energy Modelling System (NEMS), an integrated model that captures interactions of economic changes and energy supply, demand and prices. Figure 18.1 shows an example of the projections made in the AEO. It is the baseline scenario of energy consumption by fuel type from 1990 to 2050. A clear transition towards cleaner fuels is evident from the projection. The consumption of coal is expected to decrease gradually, while that of other renewables and natural gas is expected to increase gradually.

Information on the consumption of fuels is gathered and recorded in The Short Term Energy Outlook (STEO), which is a publication that is released every month. Figure 18.2 shows an example of the content of the STEO. It projects the price of different types of gasoline and crude oil until the period 2020.

Further, the International Energy Outlook (IEO) presents a detailed global energy outlook. Figure 18.3 shows an example of the content. It projects the energy consumption per capita for six countries for the period 2015–2040.

In addition to the EIA, the United States also has a national laboratory, the Pacific Northwest National Laboratory, that undertakes integrated energy modelling by connecting it to related sectors like land, water, and energy. This laboratory looks into four broad areas: physics and computational sciences, earth and biological sciences, energy and environment, and national security. It houses 4,414 scientists, engineers, and professional staff; has developed 2700 foreign patents since 1965; and has published 1,124 peer-reviewed articles in the year 2018 alone.

18.1.2 Status of Energy Modelling in India

Energy modelling happens in a fragmented way across some government ministries and several research and government organisations. This fragmented approach to energy modelling has

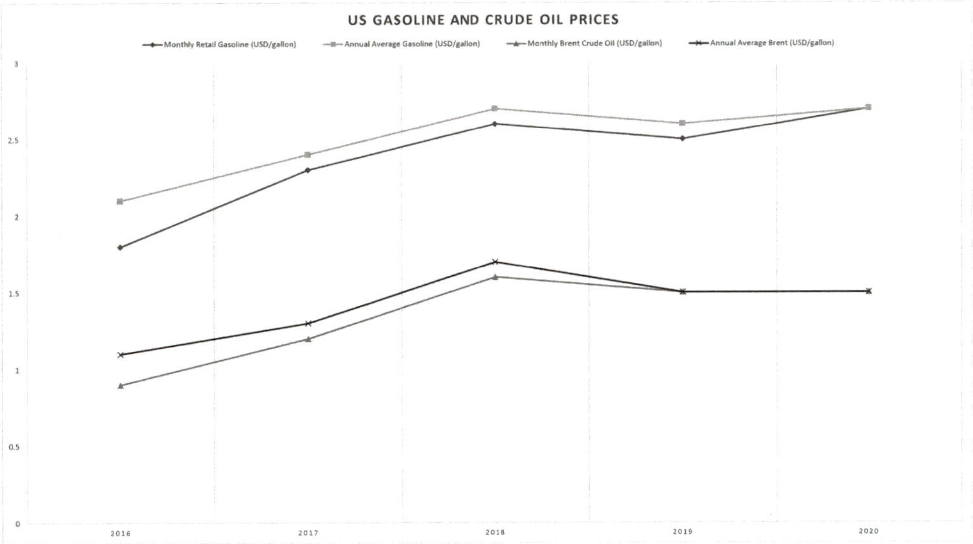

FIGURE 18.2 US gasoline and crude oil prices (data: Short-Term Energy Outlook).

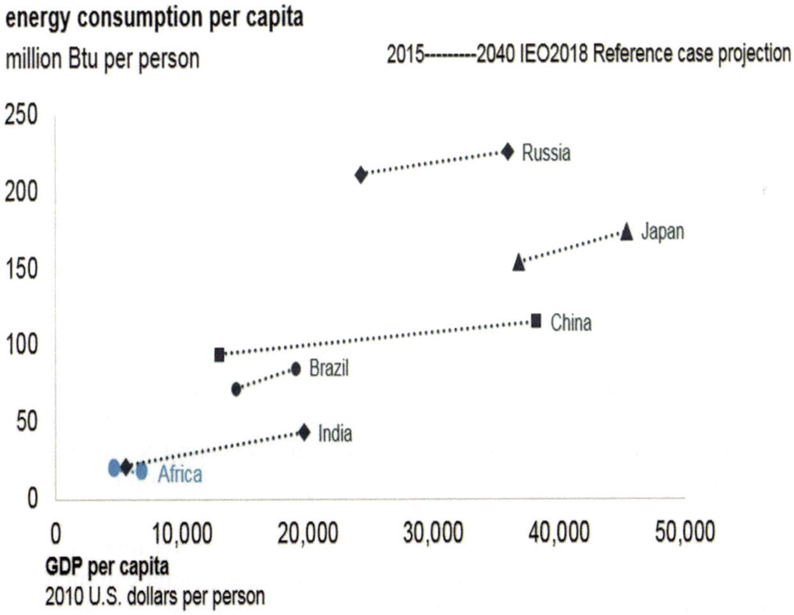

FIGURE 18.3 Energy consumption per capita (data: International Energy Outlook).

limitations. There is no government body in India that is legally mandated to undertake modelling for the entire energy sector. The Central Electricity Authority (CEA) and Petroleum Planning and Analysis Cell (PPAC) undertake modelling in house only for their respective areas and on select policy issues concerning their sector. There is a lack of integrated research in the domain of energy. Because separate research divisions under each ministry undertake technical work in their respective areas, situations inadvertently arise when sometimes ministries end up promoting their fuel over the fuels of other ministries.

The majority of energy modelling work is outsourced by the government to research organisations, academic institutions, or private consultancies. It is also need based and not regular. Some organisations which are actively engaged in this work in India are the Council for Energy Environment and Water (CEEW); Energy and Resources Institute (TERI); Center for Study of Science, Technology and Policy (CSTEP); Integrated Research and Action for Development (IRADe); and Observer Research Foundation (ORF). Along with them, there are other organisations like ICF, Shakti Foundation, World Resources Institute (WRI), Energy Policy Institute at the University of Chicago (EPIC), Rocky Mountain Institute (RMI), Prayas Energy Group (PEG), and more recently the Celestial Earth (TCE). In addition, there are also private consultancies like McKinsey, KPMG, PricewaterhouseCoopers (PWC), and academic institutions like the Indian Institute of Technology (IIT) and Indian Institute of Management (IIM) to whom work is outsourced.[1] These are need-based collaborations for informing decisions in the government.

18.1.3 Steps towards Developing a National Energy Model at NITI Aayog

In 2017, NITI Aayog took steps to develop an energy optimisation model in-house with technical support from the International Institute of Applied System Analysis (IIASA) under the umbrella of the Technology Information Forecasting and Assessment Council's (TIFAC) membership in IIASA. The model built by NITI Aayog uses the $MESSAGE_{ix}$ optimisation model of IIASA (Thambi et al., 2018). MESSAGE can be expanded as "model for energy supply strategy alternatives and their general environmental impacts". It is worthwhile to discuss some of the results from the model to demonstrate how results from such models can feed into policy making. Scenario assessments have been used extensively to inform energy and climate policy worldwide. Scenarios can be created with combinations of different types of assumptions, for instance, higher economic growth or higher renewable energy penetration.

The results from a scenario assessment of the model are presented after a description of the two scenarios.

Baseline Scenario: The baseline scenario takes into account energy and climate-related policies of the government of India, which have already been implemented. It shows a future deemed most achievable by the implementation of current policies and programmes. It assumes gradual improvement in end-use energy management in all sectors as a result of technological improvement. This scenario provides the basis for assessing alternative policy impacts.

NDC Scenario: India submitted its Nationally Determined Contribution (NDC) to the UNFCCC in October 2015, which became a part of the Paris Agreement signed by countries. NDC represents the commitment of each member country to reduce emissions and adapt to climate change (UNFCCC, 2015). As a part of India's domestic targets towards meeting NDCs, the government has set the target of achieving 175 GW of renewable energy by 2022, with 100 GW coming from solar and 60 GW from wind.

A scenario assessment between baseline and NDC scenario shows that by achieving its targets for increasing renewable capacity per the NDC commitments, India can increase the share of renewables in power generation to 22% in 2032 as compared to almost 16% in the baseline, seen in the Figure 18.4. Further, the share of non-fossil sources in installed capacity is 45% in 2030, as compared to 35% in the baseline scenario.

Energy models can be used to evaluate the effect of numerous scenarios like this to demonstrate the effect of sector-specific national policies. Further, they can be integrated with land or water models to understand the costs and benefits of a national policy across sectors. However, it is very important to understand that model results are not predictions. Their purpose is to develop scenarios based on national policies to reduce uncertainty about the future or to understand uncertainty better. To get this point across to government officials is a challenge that most modellers face when presenting results.

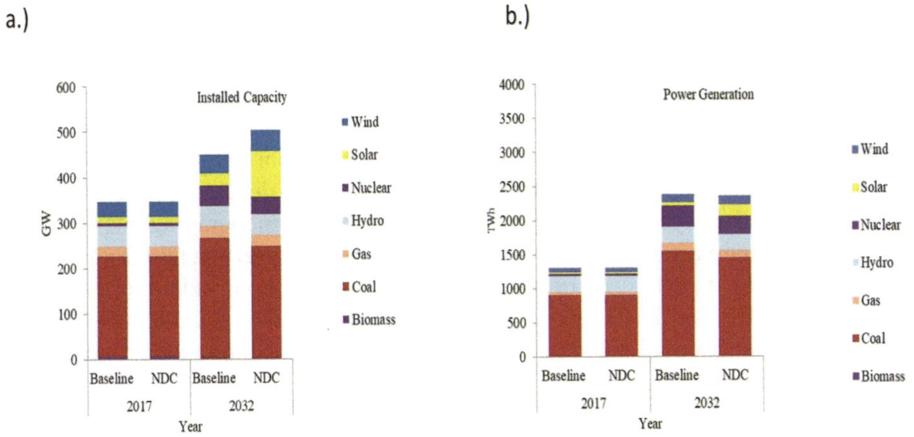

FIGURE 18.4 Assessment of NDC scenario and baseline. (a) Total installed capacity in GW; (b) electricity generation by fuel in TWh.

18.1.4 Developing a State Energy Model

For India, as a large country with an unprecedented level of diversity in its economic, social, and environmental conditions across the country, it is important to have a sub-national–level assessment in terms of its energy supply, demand, and use. States are diversified in nature and have different economic structures and thus warrant separate assessments of their economic growth and development fuelled by varieties of natural resources. Currently, there are two different types of state-level energy modelling and assessment work going on. One, state-level energy calculators are developed in line with the national-level energy calculator, called IESS 2047 (India Energy Security Pathway Calculator 2047), which was developed and has been maintained by NITI Aayog since 2012, and two, state-level energy models developed in line with the national-level India Energy Model by downscaling it to the state level. These models are primarily least-cost optimisation models which minimise the cost of the overall energy supply in the state system by fulfilling all other given constraints in the system, which could include but are not limited to availability of energy resources, renewable energy resources, land and water availability, and emissions constraints.

State-level energy calculators are primarily energy accounting tools which account for state-level energy supply, demand, and their corresponding costs and emissions. These tools are excellent sources of consolidated state-level energy data and information which could be used for multiple purposes in the future, especially developing model usable databases. These tools can also be used for state-level energy planning and policy making activities. However, the major drawback of these tools is that they cannot provide optimal solutions. Currently (as of 2018), seven state energy calculators are available in India for seven different states which are developed by different agencies in India under financial support from multilateral and bilateral agencies.

State-level energy models are primarily energy systems optimisation models which are capable of providing a cost optimal energy supply and demand portfolio at the state level given varieties of constraints. Currently, under bilateral support, one state of India[2] is developing its state energy action plan based on the state energy model using the MESSAGEix open source integrated energy modelling framework. Given the increasing complexity of the state economies in India and their interdependence with a variety of natural resources, national and international commitments, and targets, it is imperative to have optimisation models in place. There is an increasing level of interest in India from states in terms of developing state energy models and the use of these models for state planning and policy making.

18.1.5 Inter-Model Comparison Exercise

Prominent international reports like IPCC reports always base their policy implications on the findings of a handful of comparable models. Using more than one model to analyse the same issue helps in making the findings more robust. Also, given the different technical features of the models, different models have different strengths and weaknesses. It is important to leverage them meaningfully by asking the right policy questions. Recognising the need to develop expertise to use multiple models for research and analysis, NITI Aayog took a step towards encouraging inter-model comparison exercises. Under the Sustainable Growth Working Group of government of India, NITI Aayog, for the first time, provided a platform for inter-model comparison. Four Indian research institutions (CSTEP, CEEW, IRADe, and TERI) and the Pacific Northwest National Laboratory (PNNL) of the US Department of Energy coordinated to provide analysis on topics of national interest, the water-energy nexus and transportation sector.

- Study on Water Energy Nexus: The study shows that there is a reduction in water withdrawal when the water conservation rules of MoEFCC are effectively implemented in thermal power plants. A decline in water withdrawal occurs due to switching to efficient cooling tower technology and shift in share of power capacity from fossil fuels to renewable energy (Srinivasan et al., 2018).
- Study to Assess Mitigation Impact of Transport Sector Decarbonisation: The results of this study show that increased fuel efficiency and a modal shift towards public transportation will have a high potential for CO_2 emissions reduction (Paladugula et al., 2018).

All models rely on a base year, which is the starting point of every modelling study. Based on this baseline, scenarios are created for the future, which are called alternative scenarios that model implications of different policies by making changes to baseline cases using different assumptions. An inter-model comparison exercise must start with a credible starting point against which implications of alternative scenarios can be modelled. However, sometimes this task is not easy. Sometimes, base year numbers cannot be harmonised because differences in base year data may reflect a legitimate source of uncertainty. This was one big challenge faced by modellers in explaining their results to the government on the issue of harmonising a certain data source.

Through the NITI Aayog inter-model comparison exercise, it came to light that baseline numbers are particularly difficult to harmonise with Indian transport demand. Based on an analysis of different sources, including ministry publications and datasets of international organisations, studies find differences in aggregate estimates of energy consumption of the transportation sector to be around 23%. This is due to differences in sources and interpretation of existing data. In addition, there are large differences in energy consumption by mode of transport and differences in energy consumed by passenger vehicles in total energy demand in the range of 38–68% in 2010.[3] Figure 18.5 shows the differences in service demand, and Figure 18.6 shows the final demand for the transport sector by comparing various sources considered by the modelling teams under the inter-model comparison exercise. This kind of divergence in historical data of consumption has been observed in other developing countries as well, for instance, in China (Yin et al., 2015) and Russia (Kholod et al., 2016). Chaturvedi et al. (2012) papers highlight the point that differences in historical data across different data sources may not be because of poor craftsmanship of the modellers but rather due to legitimate reasons. Artificially forcing the model to harmonise data points of a specified source would not be the right approach, as it will only end up giving a fake sense of uncertainty which is not present in the real world.

To overcome the fragmented approach to energy modelling, the government of India can create an institution that integrates the various kinds of modelling initiatives within the country and undertakes energy modelling for different types of sectors collectively. This institution should encourage in-house capabilities as well, as this will help policy makers to fully appreciate and understand the

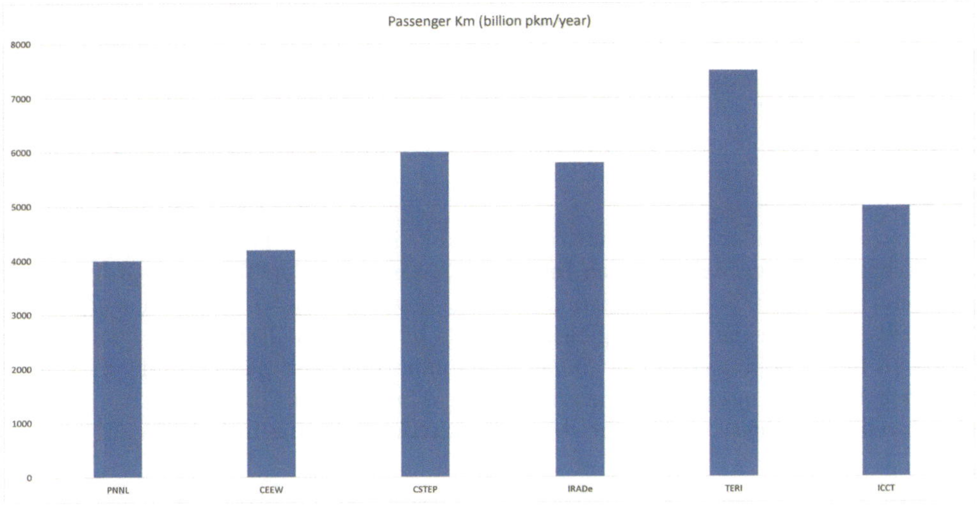

FIGURE 18.5 Total service demand for transport sector (data: Paladugula et al., 2018).

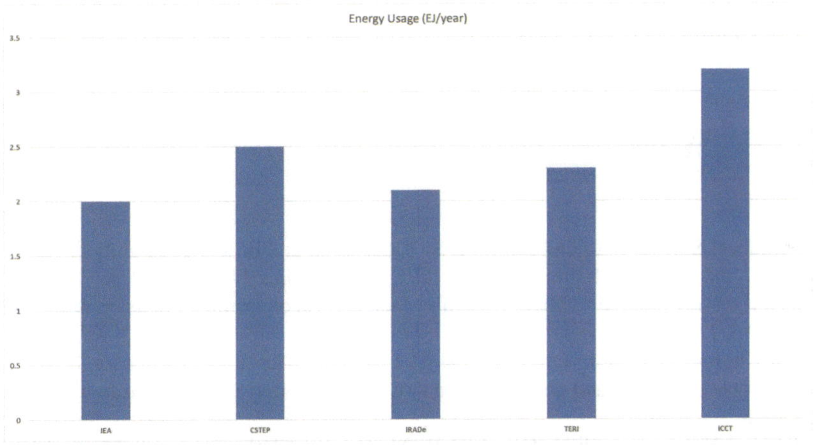

FIGURE 18.6 Final usage for transport sector (data: Paladugula et al., 2018).

challenges and opportunities and the simplicity or complexity of these models. When the government undertakes modelling, technical outputs have a greater ability to seep into the planning system. This will also benefit the modelling community because modellers sometimes face challenges in explaining the strengths and weaknesses of their technical outputs to government officials; such situations can be minimised when the officials are already aware of some technical features of these models.

18.2 IMPACT

Energy modelling for the entire energy sector should be integrated into the planning process in a sustainable, periodic, and timely manner as a part of the normal organisational functioning of national government. This will result in publication of reports based on results of modelling exercises that feed into the planning process from time to time, as we saw in the case of EIA. This will also facilitate free-flowing interaction between the models, modellers, and policy makers to encourage development of models better suited to depict Indian realities. It will build capabilities within the government to refine existing models to answer new policy questions.

The institution should conduct integrated energy modelling, that is, analysis for the entire energy sector, as well as other interconnected sectors for integrated nexus analysis, for instance, the nexus between land use in the power sector and agriculture or nexus of water use in agriculture and the power sector.

With the upsurge of inter-model comparison exercises in different parts of the world, the global modelling community has developed an Integrated Assessment Modelling Consortium (IAMC)[4] that facilitates a platform for standardisation of results across the models of different organisations. As India builds its domestic modelling base, it would be useful to align to such common modelling protocols by developing common tabular and graphical formats to ensure smooth comparison across different types of models. The government could facilitate this process.

(The chapter was written based the state of modelling activities in India in 2020.)

NOTES

1 The names mentioned here are non-exhaustive.
2 With generous funding from GiZ, the state of West Bengal is implementing the state energy plan and action plan using the state energy model (least-cost optimization model on MESSAGEix open source platform). The Celestial Earth and KPMG are jointly implementing this pioneering work in India at a state level.
3 Paladugula et al. (2018).
4 The Integrated Assessment Modelling Consortium is an organisation of scientific research organisations that pursues scientific understanding of issues associated with integrated assessment modelling and analysis.

REFERENCES

Chaturvedi V., Waldhoff S., Clarke L. and Fujimori S. 2012. What are the starting points? Evaluating base year assumptions in the Asian modelling exercise. *Energy Economics* 34, S261–S271.

Kholod N., Evans M. and Kuklinski T. 2016. Russia's black carbon emissions: Focus on diesel sources. *Atmos. Chem. Phys.* 16, 11267–11281.

Paladugula A. L., Kholod N., Chaturvedi V., Ghosh P. P., Pal S., Clarke L. and Wilson S. A. 2018. A multi-model assessment of energy and emissions for India's transportation sector through 2050. *Energy Policy* 116(C), 10–18.

Srinivasan S., Kholod N., Chaturvedi V., Ghosh P. P., Mathur R., Clarke L., Evans M. et al. 2018. Water for electricity in India: A multi-model study of future challenges and linkages to climate change mitigation. *Appl. Energy*210, 673–684.

Thambi S., Bhatacharya A. and Fricko O. 2018. *India's Energy and Emissions Outlook: Results from India Energy Model*. NITI Aayog.

UNFCCC. 2015. India's intended nationally determined contribution: Working towards climate justice. Available at: https://unfccc.int/sites/default/files/NDC/2022-06/INDIA%20INDC%20TO%20UNFCCC.pdf

Yin X., Chen W., Eom J., Clarke L. E., Kim S. H., Patel P. L., Yu S. and Kyle G. P. 2015. China's transportation energy consumption and CO2 emissions from a global perspective. *Energy Policy* 82, 233–248.

19 Evaluation Study to Assess the Performance of Sanitation Coverage in Selected States

Prerna Mukharya

ABSTRACT

The government of India prioritised sanitation and hygiene with various campaigns launched across the country in the early 1990s. Today, the sanitation issue has been revived with much vigour under the Swacch Bharat Abhiyan, an initiative of the central government. This study aimed at evaluating the trends in sanitation and hygiene in five states that had performed above average concerning sanitation coverage, and data-based insights and recommendations were provided.

The study analyses the trends in sanitation and identifies the inflection points for policy change, local factors, agencies, and other change-makers in the field of sanitation. The study then explores the primary reasons behind what, how, and why the major changes in the sanitation sector occurred in the five states and analyses the impact on health indicators.

19.1 INTRODUCTION

Sanitation and hygiene are a national priority. It is the Gandhian way of life; therefore, sanitation has been a national objective since pre-independence. To this end, a plethora of campaigns have been launched across the country since the early 1990s. In the present, sanitation has gained even more importance ever since the Swacch Bharat Abhiyan, a central government initiative, has swept the nation.

The Swachh Bharat Abhiyan (Clean India Mission) is a nationwide campaign launched by the government of India on October 2, 2014, to honour Mahatma Gandhi's vision of a clean India. The mission aims to eliminate open defecation, improve solid waste management, and create open defecation–free (ODF) villages.

Data available at the Swachh Bharat Mission (SBM) portal as of January 2015 indicates that a few states have performed distinctly better than others in the field of sanitation. The top five states as identified are Sikkim, Himachal Pradesh, Manipur, Meghalaya, and Kerala. The present study explored the motivating factors that were instrumental in the success of the sanitation policies across these states.

19.2 CURRENT SCENARIO

The present study evaluated the trends in sanitation and hygiene in five states that had performed above average for sanitation coverage, per the Ministry of Drinking Water and Sanitation (MDWS) 2012 baseline survey. The main objectives of the study were to first analyse trends in sanitation and identify the inflexion points with respect to policy change, local factors, agencies, and other change makers in the field of sanitation. Second, we wanted to explore the primary reasons behind what, how, and why the major changes in the sanitation sector occurred in the five states and analyse the impact on health indicators.

DOI: 10.1201/9781003487531-24

Our objective was also to identify the major challenges faced by officials and agencies involved in sanitation campaigns. Finally, we wanted to make recommendations for better implementation of sanitation policies at the state and national level.

19.2.1 Methodology

We used specific methodologies to guide the investigation. We compiled secondary data on sanitation coverage and related health indicators for each of the five states. We also reviewed existing literature, including government reports, reports from donor agencies, articles, and other communication materials to get a better idea of the status of sanitation in the five states. Overall, an ethnographic approach was adopted, and the prepared tool was fairly open ended. The basic structure of the tool was designed based on the literature review, keeping in mind the broad theme of the objectives.

A purposive sampling method was followed to shortlist individuals and organisations (government and non-governmental) in the five states. These individuals were contacted and interviewed personally and telephonically. Focused group discussions were also conducted with staff or members of relevant bodies, like village sanitation committees, gram sabhas, NGOs operating in the field, and other community members. Villages and towns where sanitation policies had been implemented or that demonstrated high levels of sanitation target achievements were visited.

The study team interacted with 30 individuals from more than 15 government and non-government organisations across the five states. Ten different gram panchayats (GPs) were visited, and local community members, including gram sabha leaders, ward leaders, angadwadi workers, and others, were involved in the discussions. This provided a varied and comprehensive idea of key factors responsible for successful sanitation campaigns. The key findings from the study are listed in the following

19.2.2 Case Studies[1]

19.2.2.1 Sikkim

In February 2014, the fund utilisation against release was nearly 99.2% for sanitation. Under the Nirmal Bharat Abhiyan (NBA) scheme, 100,798 individual household latrines were constructed against the target of 87,014 to January 2014, achieving 115.84% of the target. Massive forest cover

FIGURE 19.1 "Green Sikkim" highlighted on the roadside.

and many spring water sources are other factors that act as enablers. Sanitation was linked to spring protection and thereby to drinking water security.

In Chuba Phong GP, water was a problem. A proposal was submitted at the gram sabha level to construct a tank for each household. "This would help in maintaining cleanliness and could be used in the kitchen garden as well. More importantly, it could be used for farming in the off-season", said Mr. Milan Kumar Chettri, panchayat member, Chuba Phong GP. He is considered instrumental to the project, which led to a National Award for exemplary work done under the Mahatma Gandhi National Rural Employment Guarantee Act (MGNREGA) for the GP. The project was initially set up in Namthang but has since been extended to other blocks and districts across Sikkim.

Dr. Chamling, ex-chief minister, launched a community-led total sanitation campaign in 1999 for achieving full sanitation across the 7,096 sq. km area of the state. The state government instituted a Nirmal Gram Puruskar (Clean Village Award) to encourage sanitation.

Between 2003 and 2005, the government undertook various initiatives, and by 2008, the results began to become obvious. According to Chewang Lachenpa, general secretary, Lachen Tourism Development Committee (LTDC), "The Sanitation Cell at Rural Management and Development Department (RMDD) was not so active earlier. After 2008, the Cell became pro-active".

Under the rural water and sanitation scheme, all households have been sanctioned a tank, which is to be funded by the government. As mentioned earlier, 90% of the fund for toilet construction is provided by the government. Ten percent has to be contributed by the beneficiaries. NGOs are responsible for providing human resources.

"Under MGNREGA, where there is water scarcity, we provide water-harvesting tanks. Rainwater is also harvested in the rainy season and used for irrigation in the dry season", points out Mr. Shital Pd. Pradhan, chief engineer, RMDD, government of Sikkim.

Supervision in the field has been a key to good implementation practices in Sikkim. Officers and panchayat members make frequent field visits to understand ground realities, connect with the people, and devise innovative methods for improvement. For instance, in Namthang block, South Sikkim district, A.B. Gurung, the BDO officer, set up whiteboards in common areas to gather public opinion and complaints.

"Grazing is not allowed in forests; only in stipulated areas", says an engineer at the RMDD. As a result, greenery has improved over the years. In 1998, the government banned grazing in reserved forest areas, plantation areas, and water source areas. However, this ban was made stringent only in the last few years.

Every GP has a member called a "barefoot engineer" who is responsible for identifying proposed works, conducting technical surveys, and supervising masons. "They are given machines for water treatment and the water is tested every quarter. They are also responsible for repair work", says Mr Zion Lepcha, secretary, Nagi Karek GP. He adds that if the funds for repair fall short, a local collection drive is organised or the ward funds salaries from households on a volunteer basis. The use of plastic has also been banned in the state.

To instil awareness among school children, the issue of sanitation could be incorporated into the school curriculum. As a step forward, the government has sent content to the HRD ministry that could be converted into chapters for school textbooks. "The government recognises that toilets for girls are necessary to reduce dropout rates and hence this is a priority issue", says Mr. D.R. Nepal. He further adds, "At the senior secondary level, sanitary napkins are provided and vending machines have been installed".

To counter the problem of water contamination, two state-level laboratories have been set up in South and East Sikkim to test water. We were told that water in Sikkim does not have chemical contamination, only a certain amount of bacterial contamination.

A sanitation programme is organised every Saturday, wherein the panchayat secretary and members are present to motivate people and campaign. For over 10 years, "cleanliness drives" have been organised in Sikkim where streets are swept and paper and plastic are collected and disposed

FIGURE 19.2 Waste collection in Sikkim.

of. Recently, this has been given a name under the PM's campaign of Swachh Bharat Mission, but Sikkim has had such practices for a long time.

Twice a year, a Panchayat Sammelan is held, and among other subjects, health and sanitation are also discussed. Praising the CM's efforts, Mr. Shital Pradhan added that,

> Sanitation issues need to be addressed on continuous basis, as the population levels are always in a state of flux. Even though total sanitation was achieved in the state by 2008, new households that have come up needed to be covered as well. Therefore, at all such events, the CM emphasizes the need to have 100 percent sanitation. The same goes for schools, wherein with more enrolments, we increase the number of units.

He constantly discussed with the community since solid waste management compost centres have been set up all over Sikkim. The one in Namthang was the first, established in 2010, where non-degradable/plastic items are recycled to make sellable items such as dolls, cushions, and pots for planting saplings.

The government mandates that all candidates contesting MLA and panchayat elections must have toilets in their house to be eligible. Nagi Karek, Namthang block, has set up a Facebook page to discuss local issues and reach out to youngsters. Also, a website was set up which is currently inactive.

19.2.2.2 Meghalaya

Mawlynnong village: "the cleanest village in Asia – a marker of achievement" (East Khasi Hills district, Meghalaya). It was the recipient of the Nirmal Gram Puruskar in 2010. This is a Khasi-Christian matrilineal village with 100% literacy. It was touted to be the cleanest village in Asia by *Discover India* magazine in 2003, the BBC in 2005, and the Meghalaya Tourism Department. Recently, it has also become a tourist destination. It is described as a "community-based ecotourism initiative".

The following are the reasons this place could be made into a tourist destination:

- It is close to the Bangladesh border
- There are two living root bridges in a neighbouring village called Riwai

Initiatives by the community to maintain hygiene: Bamboo baskets are kept at 100 m to collect solid waste. The waste is then disposed of by the villagers on a voluntary basis, while the kitchen waste is used as manure. The rest of the waste is used as fuel or for fire, and other non-biodegradable waste (although plastic is completely banned in the village) is disposed of in the forest.

Liquid waste is disposed of through the typical nala system, which is common in the north east and pre-dates the modern drainage system, which is absent in Meghalaya. Water does not stagnate since the area is sloping.

There is no lapse in cleanliness throughout the day, and duties are clearly assigned. There are numerous signboards that ask visitors to keep the area clean. Children are assigned the task of cleaning the area around their house as soon as they reach class 1 and 2. Failing to do so means they do not get their meal.

It is reported that the village started following the cleanliness regime long before any national/state sanitation intervention was done. It is claimed that someone from the village had the chance to go abroad, where he learnt the importance and significance of cleanliness and hygiene. He mobilised the village and initiated a systematic cleanliness drive as a necessary step to foster tourism. Currently, NGOs such as the Bethany Society are also imparting skill training to the villagers to help them make indigenous products to sell as local souvenirs instead of buying goods from Bangladesh and selling them as their own. These measures have positively impacted the people's economic lives in the village.

19.2.2.3 Mawsynram Block, East Khasi District, Meghalaya

Potentially Mawsynram area is deemed the rainiest place in the world. It overlooks Cherrapunjee and Bangladesh through a bed of clouds. It also has the Langsymphut peak, waterfalls, and a hot spring, which could make it a tourist destination. Currently, it has poor roadways. Improvement of its roads could bolster its economy. The maintenance of hygiene and the title of an ODF village are perceived to be a step forward in this direction, leading to the economic advantage that the community hopes to garner from tourism.

Measures undertaken in a particular cluster, 4 out of 15 villages (Ponkung, Mawhiang, Wahmawpat, and Langsymphut) have been declared open defecation free, with the fifth lined up. The declaration of ODF is seen as a symbol of prestige, since open defecation is conceived of as a stigma. The "declaration" day is popularly called the "wedding day of the village", said Mr. Mark West, district coordinator.

The threefold process is verification, certification, and celebration, which involves a grand celebration with "community members dressing up in good clothes", said Mr. West. Mawhiang was the first village to be declared ODF in June 2014, and that became a motivation for the surrounding villages that are located close to each other.

Innovative practices or indigenous microorganisms (IMOs) are commonly used by the community in Mawhiang and its neighbouring villages. This was introduced by the Bethany Society (the district being a project area of the NGO) and its senior programme coordinator, Mr. Ricky Renthalei, who learnt it during his visit to Japan. IMOs are used as a natural sanitising and deodorising agent prepared as a mixture of rice bran and jaggery, which is moistened and kept for fermentation in a pit that is dug in the forest area. The mixture is then put around pigsties (livestock).

19.2.3 Key Findings

1. *Community-based action:* All five states have involved community leaders, organisations, and agencies in spearheading the sanitation drive. These may include panchayat representatives, anganwadi workers, ward leaders, and swachchta doots/workers. Participants

unanimously emphasised the role of community representatives in spreading awareness and mobilising people toward adopting good practices. Grass-root mobilisation, community engagement, and awareness-building programs were exponentially more effective in the sanitation campaigns across these states vis-à-vis target-based subsidy policies.

2. *Government efficiency*: The takeaway from states like Sikkim and Himachal Pradesh is the instrumental role of smooth governmental functioning and dedicated contribution by state-level bodies towards encouraging and supporting community-based campaigns. Attention towards coordination and establishing a clear line of communication was key. This is not only in the transfer of funds but also in how government bodies respond to and consider the needs and concerns of different stakeholders, conduct regular review meetings, and take timely action on issues identified.
3. *Panchayati raj and village councils*: Most of the states in the study have largely involved village administration and panchayat bodies in the dispersal of implementation policies. Barring Sikkim, it is only local leaders and representatives who can successfully reach out to their constituencies and villages and effectively supervise the transformation that these campaigns entail.
4. *External agencies*: States like Kerala, Sikkim, and Himachal Pradesh received substantial support from external agencies, whether national NGOs or international organisations like World Bank and WWF, who provided the requisite expertise and momentum to these campaigns. Such efforts did contribute to these states overachieving on set targets.
5. *Innovative practices*: Some of the states have devised innovative local technologies and practices such as toilets made of tin sheets and bamboo in Manipur, IMOs and ecological sanitation (ECOSAN) in Meghalaya, and the worming pit in Himachal, which are ingenious ways of dealing with local needs within the possibilities available to the people.
6. *The role of women:* One of the primary findings from the report is the contribution of women workers and leaders who have been identified in most states as essential change makers in the sanitation and health department. It is therefore recommended that other states and national campaigns incorporate this aspect and involve a larger number of women activists and workers in future projects.
7. *Cultural and social practices:* A wide range of cultural and social practices were motivators for the success in these states. This includes indigenous practices, often conceptualised and regulated by village councils and villagers themselves like the market day/village haat in Meghalaya. Moreover, the people of Sikkim are culturally tuned towards hygienic sanitation practices, while in Kerala, the development of sociopolitical consciousness played an essential role in creating awareness on health and sanitation.
8. *Incorporating sanitation within the ambit of good health and social standards:* Some of these states' success stories have involved incorporating sanitation within the general idea of good health. Often, people were disinterested in changing sanitation habits until its health benefits were discussed, along with possible implications for social status and community acceptance.
    - The main challenges today are for management of solid and semi-solid waste, especially in and around rural and urban areas.
    - In low-lying coastal terrains like Kerala, the issue of construction of safe toilets remains, since the area is low lying, posing a threat to groundwater quality and making it additionally difficult to construct pits.
    - In the mountainous terrains of Sikkim, Himachal, Meghalaya, and Manipur, connecting pipelines from the water source to the village or individual households is difficult. Inaccessibility to water for cleaning purposes induces open defecation.
    - While the hope for tourism has encouraged sanitation in Meghalaya, in Himachal and Sikkim, which are popular destinations, tourism has contributed to the problems

of waste management and cleanliness. Certain states have made concerted efforts to combat the cleanliness challenges.
- The findings from Manipur point to the need to pay immediate attention to regions and states suffering from issues of security and political disturbances. This has not only affected access to water sources but also artificially raised its price. In such conditions when basic necessities are not guaranteed, campaigns will not achieve their proposed targets.
- In some of the states, while individual latrine construction and use has now reached an optimum level, community complexes, for instance, in schools and markets, have not been successful, primarily because of the lack of supervision. Sustainability of these efforts is therefore crucial and must be discussed at a macro level.

9. Last, a general force behind the achievement in these states is the overall good status of education, income, access to resources, and the consequent consciousness regarding health and hygiene and a concerted and coordinated effort by the government machinery, especially in the smaller states.

19.3 IMPACT

The five states included in the study provide a varied and comprehensive idea of the key factors responsible for making sanitation campaigns successful. The following are the enablers.

19.3.1 Community-Based Action

All five states have involved community leaders, organisations, and agencies to spearhead the sanitation drive. These may include panchayat representatives, anganwadi workers, ward leaders and swachhata doots. In fact, participants unanimously emphasised the role of community representatives in spreading awareness and mobilising people towards the adoption of good practices.

19.3.2 Government Efficiency

The "take away" from states like Sikkim and Himachal Pradesh is the instrumental role of smooth governmental functioning and dedicated contribution by state-level bodies towards encouraging and supporting community-based campaigns. This is not only in the form of transfer of funds but also in how government bodies respond to and consider the needs and concerns of different stakeholders.

19.3.3 Panchayati Raj and Village Councils

Most of the states in the study have largely involved village administration and panchayat bodies in the dispersal of implementation policies. Hence, strengthening and supporting local governmental bodies is a primary step towards ensuring that policies regarding individual household-based changes, like that of sanitation, are implemented. Barring Sikkim, it is only local leaders and representatives who were able to successfully reach out to their constituencies and villages and effectively supervise the transformation that these campaigns entail.

19.3.4 External Agencies

States like Kerala, Sikkim, and Himachal Pradesh received substantial support from external agencies, whether national NGOs or international organisations like World Bank and World Wildlife Fund (WWF), who provided the requisite expertise and momentum to these campaigns.

19.3.5 Innovative Practices

Some of the states have devised innovative local technologies and practices like toilets made of tin and bamboo in Manipur, IMO and Ecological Sanitation in Meghalaya, and the worming pit in Himachal, which are ingenuous ways of dealing with local needs within the possibilities available to the people. This approach is recommended for other states as well, where organisations and stakeholders need to think of low-cost and effective ways of improving sanitation practices.

19.3.6 The Role of Women

One of the primary findings from the report is the contribution of women workers and leaders, who have been identified in most of states as essential change makers in the sanitation and health department. It is, therefore, recommended that other states and national campaigns incorporate this aspect and involve a larger number of women activists and workers in future projects.

19.3.7 Cultural and Social Practices

A wide range of cultural and social practices were motivators for the success in these states. These include locally based practices, often conceptualised and regulated by village councils and villagers, like the market day in Meghalaya. Moreover, the people of Sikkim are culturally tuned towards hygienic sanitation practices, while in Kerala, the development of socio-political consciousness plays an essential role in creating awareness regarding health and sanitation.

19.3.8 Incorporating Sanitation within the Overall Idea of Good Health and Social Standard

Some of these states' success stories have involved incorporating sanitation within the general idea of good health. Often, people were disinterested in changing their sanitation habits until its health benefits were discussed, along with possible implications for social status and community acceptance.

Finally, a general force behind the achievement in these states is the overall good status of education, income, access to resources, and the consequent consciousness regarding health and hygiene. Therefore, if similar statistics are to be achieved nationally or in other states, the general state of education and access to resources need to be improved.

NOTE

1 Drawn from Outline India's report titled *Evaluation Study* to assess the performance of five above-average states in sanitation coverage.

BIBLIOGRAPHY

Dash, Amarendra Kumar, and Rajendra Kumar Dash. "Environmental and sustainability campaigns: A case study of India's Swachh Bharat Abhiyan (2014–2019)." *Journal of Communication Management* 25, no. 4 (2021): 385–400.

Jangra, Babita, J. P. Majra, and Mahavir Singh. "Swachh Bharat Abhiyan (clean India mission): SWOT analysis." *International Journal of Community Medicine and Public Health* 3, no. 12 (2016): 3285–3290.

20 Data-Driven Approach to Determine Distribution Centres in Pradhan Mantri Ujjwala Yojana

Prukalpa Sankar and Richa Verma

ABSTRACT

Hon'ble Prime Minister Shri Narendra Modi launched the Pradhan Mantri Ujjwala Yojana (PMUY) on May 1, 2016. Through this scheme, the government aims to provide LPG connections to below poverty line (BPL) households in the country. Ten thousand new LPG distribution centres had to be opened in locations for almost every citizen in each of the 6.4 lakh villages to reach a distributor within 10–15 km of his/her residence. The chapter discusses the data science–based approach for deciding the ideal locations for the new distribution centres.

The data of the currently operating LPG centres was put on a map, along with other relevant criteria such as population, LPG penetration, and income levels. Then access was made the primary factor for deciding new centre locations. The government's approach was not based on market potential but on opening centres in the vast blank of the nation's geography where people do not have access to LPG and thus have to resort to other fuels for cooking.

20.1 INTRODUCTION

The Pradhan Mantri Ujjwala Yojana (PMUY), a social welfare scheme, was launched by Hon'ble Prime Minister Shri Narendra Modi on May 1, 2016, in Ballia, Uttar Pradesh. Under the PM Ujjwala Yojana, the government aims to provide liquefied petroleum gas (LPG) connections to below poverty line (BPL) households in the country. The scheme aims to replace unclean cooking fuels mostly used in rural India with clean and more efficient LPG. Thus, it aims to safeguard the health of women and children by providing them with a clean cooking fuel, LPG, so that they don't have to compromise their health in smoky kitchens or wander in unsafe areas collecting firewood.

Under this scheme, 5 Cr LPG connections will be provided to BPL families with support of Rs. 1600 per connection in the next three years. Ensuring women's empowerment, especially in rural India, the connections will be issued in the name of the women of the households. Rs. 8000 Cr. has been allocated towards implementing the scheme. Identification of the BPL families will be done through socio-economic caste census data. However, effective implementation of the scheme required a lot of optimization and network planning.

To promote 100% LPG use in the country, data science helped to do the maths for the 10,000 best locations across far-flung areas to open new LPG dealership centres.

Village Gurundia is one of the 51,311 dots on the map of the low-income state Odisha – one for each village. The dot gets bigger on the district map – showing 30 in all – and changes into an amoeba-like shape on the map of Sundergarh, one of the state's 315 blocks. Zoom down further, and you can see the house of Sunita, a 35-year-old woman who can't help smiling these days because of

DOI: 10.1201/9781003487531-25

the ease with which she got a new LPG cylinder from an Indian Oil dealership centre a mere 4 km away. She no longer has to depend on a smoke-spewing chulha to cook for her large family.

If 2 crore women like Sunita are feeling lucky today, it is because, for the first time, LPG distribution centres in the country are being commissioned to be opened in rural and semi-urban areas within the reach of every Indian home. The aim is to specifically deliver cooking gas to 5 crore women below the poverty line who need clean cooking fuel the most.

The newly located dealership centres closer to home will help ensure that when Sunita and her friends need their next refill cylinder in two months' time, they don't opt out because of a 20-km walk saddled with a heavy cylinder. LPG customers are serviced by their respective area dealers, mostly located far off in remote areas, thanks to the limited market.

The Ministry of Petroleum and Natural Gas is spending Rs. 8,000 crore over three years to support the initial LPG installation cost by Rs. 1,600 for each BPL woman beneficiary. The women can buy gas stoves and refill cylinders in instalments. This has been one of the largest and fastest executions of a welfare scheme nationally.

20.2 INTERVENTION

In tandem with giving subsidised connections, the government is working to ramp up the LPG distribution infrastructure. In three years' time, 10,000 new LPG centres will be opened in locations for almost every citizen in each of the 6.4 lakh villages to reach a distributor within 10–15 km of his/her residence.

This is no mean task. To plan this, decision-makers needed to pore over where in the country's over 600 districts the LPG distributors were currently located. This would help them identify the exact LPG-deprived villages. "The ministry wanted five crore women beneficiaries to be identified purely on the digital platform . . . the scheme to be implemented without any human intervention", says the petroleum and natural gas minister, Dharmendra Pradhan.

"To sustain a scheme like this, you need information", he adds, "You need a monitoring mechanism which gives you the reality of the ground".

In short, you need data science, an optimization algorithm to identify the ideal locations for the new centres.

20.3 CURRENT SCENARIO

20.3.1 Urban Bias

In India, each of the 18,000 LPG distribution centres is operated by one of the three government-owned oil marketing companies – Indian Oil Corporation Limited, Hindustan Petroleum Corporation Limited, and Bharat Petroleum Corporation Limited – as independent for-profit businesses. They were given the commercial freedom to select and regulate distribution centres ever since the administrative price mechanism was dismantled in 2002.

Oil marketing companies (OMCs) selected locations for new centres based on their market potential, which led to their concentration in urban areas. Take, for instance, Lucknow district, home to the capital of Uttar Pradesh. Of the 88 LPG centres, 75 are in urban areas and only 13 in rural locations. As against an LPG distribution centre for every 7,500-odd households in the urban areas of the district, there was only one to service an estimated 21,500 homes in rural Lucknow. That was the skewed distribution the government was seeking to address.

20.4 SOLUTION

So the ministry set its sights beyond simple market economics and looked in the right place – the vast chunks of areas in the country deprived of cooking gas. The focus was specifically on the 14 lagging states and union territories.

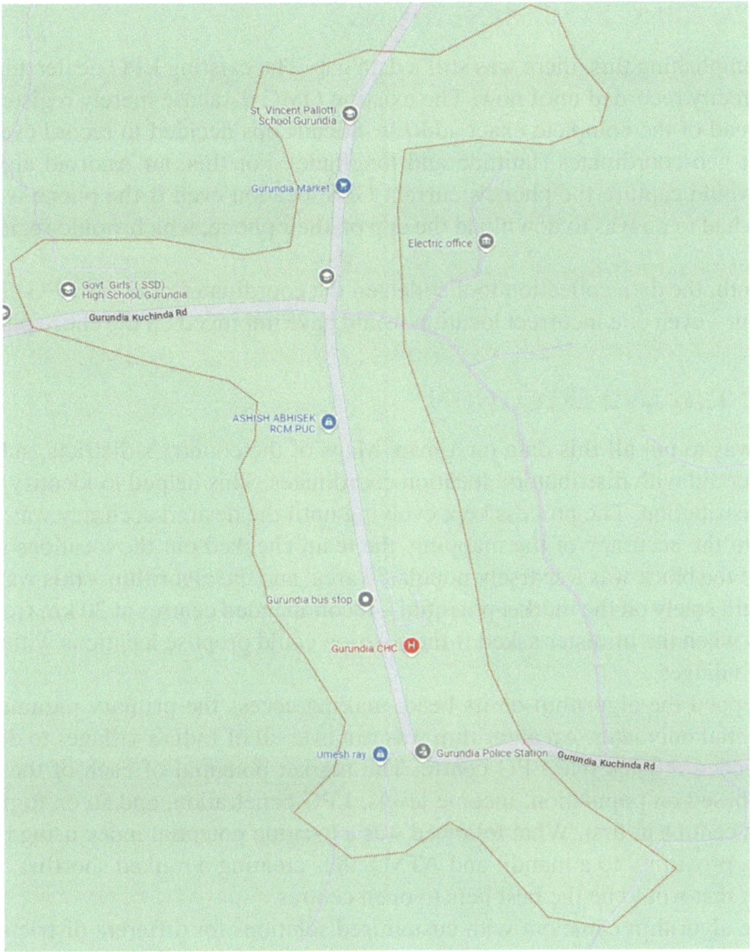

FIGURE 20.1 Map of Gurundia Village in Odisha.

There were two major problems. On the one hand, the solution had to ensure that the dealerships were viable for those setting up centres, while on the other, the new centres had to be so located as to create LPG access for the underprivileged, who had to either commute long distances for cooking gas or depend on the chulha.

There could have been two approaches to tackle this. One, the OMCs could have looked top down, identified locations based on population and market potential on a district level, and sanctioned a distribution centre. Or they could have assessed where there was an actual need through bottom-up data mapping and identified the ideal locations for new LPG centres scientifically based on an algorithm.

20.5 TYPE OF DATA USED

For the first time in the history of centre creation, market potential wasn't going to be the only criterion. For this purpose, the only data SocialCops had to get started was the refill sales data from OMCs that showed where the existing market was. But this data alone would never put an LPG centre near the homes of the poor because this data was based on separate criteria. So the market data was layered with a lot of publicly available data– on population, income levels, roads, local mandis, ATMs, post offices, health centres, and of course LPG penetration to feed the algorithm for the solution.

20.6 GEO-TAGGING EXACT LOCATIONS

Even after accomplishing this, there was still a data gap. The existing LPG dealership locations had never been correctly recorded until now. The existing OMC database merely registered the district of a dealer instead of the complete exact address. SocialCops decided to record every LPG dealer in India's exact geo-coordinates (latitude and longitude). For this, an Android app, Collect, was deployed that would capture the phone's current GPS location even if the phone were offline. All that the dealers had to do was to download the app on their phone, which would register the latitude and longitude.

In just a month, the data collection tool collected the coordinates of every LPG centre. This was a crucial element – even one incorrect location would have impinged on the entire planning process.

20.7 PLANNING AND EXECUTION

The next step was to put all this data on a map. Maps of the country's districts, sub-districts, and blocks were overlaid with distributors' location coordinates. This helped to identify areas with and without LPG distribution. The process kept evolving until the desired accuracy was reached.

Satisfied with the accuracy of the mapping, the team checked out the locations suggested. The northern part of the block was a sparsely populated area, and the algorithm – this was the first iteration that was built solely on the market potential – recommended centres at 20 km from the remotest villages. That's when the minister asked if the solution could propose locations within a reasonable distance for all villages.

The team flipped the algorithm on its head, making access the primary parameter – followed by market potential only later. An algorithm was run over all of India's villages to divide them into clusters that needed at least one LPG centre. The market potential of each of these clusters was now identified based on population, income levels, LPG penetration, and so on to prioritise which clusters to open centres in first. What followed was a location potential index using parameters like access to roads, proximity to a mandi, and ATMs, thus creating a ranked shortlist of ten locations in every district that would be the best bets to open centres.

The intuitive algorithm came out with customised solutions for different districts depending on how densely populated they were – the market potential goes higher with greater density. So, if the algorithm suggested that LPG centres be located at a distance of 15 km in Himachal, Uttarakhand, or Chhatisgarh, in UP, it was just half that distance. The solution could vary even within a state; in Gujarat, for example, centres were recommended at 7.5 km in Ahmedabad and Gandhinagar, whereas in Kachch and Narmada, it was 15 km. In other words, the algorithm balanced the distance as well as population a newly opened centre would serve.

Already data has gone on the ground, and 6,000-odd locations are being advertised for tendering to begin. In bigger states like UP, Punjab, and Orissa, the process is well underway and dealerships have begun to open.

"When we started the scheme, we were not sure if we'll be able to roll out uniformly across the country . . . whether we'll be able to achieve the numbers we are talking about", says Ashutosh Jindal, joint secretary in the Petroleum Ministry. Village-level information and solutions are not easy, he says, "But technology has been a great enabler".

20.8 IMPACT

This was a unique experiment wherein decision-makers, sitting right in Delhi, used data science to come up with a pan-India solution to address the conflicting objectives of both market and access for the remote corners of the country. "For the first time, a planning exercise that would typically take two years and suffered from subjectivity, could be tackled in a matter of months through a standardised solution", says Sankar.

The data-based solution for Ujjwala is essentially an exercise in resource allocation. It might well work as the blueprint for newer infinite use cases – to open schools, health centres, or agriculture training counters. The variables might differ and the weights might vary, but the process remains essentially the same, says the SocialCops co-founder. It can be replicated by multiple government agencies to implement their own schemes. In fact, any industry can use this blueprint for distribution now that the groundwork has been done in creating a database for these markets. Recall how the insurance industry used to keep away from the rural markets because distribution was a tough nut to crack in the initial days of deregulation?

The takeaway from the Ujjwala model is not just that there are markets beyond urban centres but that these can be tapped into profitably with data as the next 900 million Indians emerge as new consumers. That's because the cost of data-driven decisions overrides the price of guesstimates.

BIBLIOGRAPHY

Aggarwal, Shubham, Sudhanshu Kumar, and Manoj Kumar Tiwari. "Decision Support System for Pradhan Mantri Ujjwala Yojana." *Energy Policy* 118 (2018): 455–461.

Dabadge, Ashwini, Ashok Sreenivas, and Ann Josey. "What Has the Pradhan Mantri Ujjwala Yojana Achieved so Far?" *Economic and Political Weekly* (2018): 69–75.

How Big Data Can Influence Decisions That Actually Matter | Prukalpa Sankar | TEDxGateway (2017). URL: https://www.youtube.com/watch?v=C6WKt6fJiso.

Swain, Swadhina Shikha, and Pulak Mishra. "Determinants of Adoption of Cleaner Cooking Energy: Experience of the Pradhan Mantri Ujjwala Yojana in Rural Odisha, India." *Journal of Cleaner Production* 248 (2020): 119223.

21 Smart Sanitation at Kumbh 2019

Saloni Goel

ABSTRACT

Kumbh Mela is the world's largest religious festival and attracts millions of pilgrims each year. Due to astronomical footfall, the mela had become infamous for its sanitation. In 2019, the Prayagraj Kumbh Mela came up with a smart sanitation project wherein the number of toilets was increased to 80,000. To monitor these, an IoT-based smart toilet monitoring system was developed that generated real-time feedback on toilet maintenance and targeted intervention.

The system received feedback in real time, processed it, and shot off targeted messages to the responsible ground personnel. The monitoring staff would walk over to the toilets at designated time slots, log in with their credentials, and give feedback on the sanitation condition. This way, the sanitation workforce targeted only unclean toilets and saved time as well as increasing their efficiency. The Kumbh Mela of 2019 was awarded the Guinness World Record for exceptional sanitation.

As the world reflects on the triumphs of the recently concluded Mahakumbh 2025, it is worth revisiting the extraordinary transformation that unfolded six years ago in Prayagraj. The 2019 Kumbh Mela became a landmark in India's public health and sanitation history, recasting what was once a chaotic and unhygienic spectacle into the cleanest temporary city ever constructed.

The Kumbh Mela is the world's largest religious gathering, attracting millions of pilgrims. Traditionally, the mela had long been infamous for poor sanitation and hygiene conditions. However, 2019 marked a watershed. The Prayagraj Kumbh showcased an ambitious, tech-led transformation that not only reimagined the hygiene narrative but also positioned the 3200-hectare tent city as a model of cleanliness and efficiency.

An IoT-based smart sanitation system was introduced across the 80,000 toilets in the mela area. This two-way system provided real-time feedback on toilet maintenance and enabled targeted messaging to the implementation team for rapid corrective action. The result was an unprecedented level of cost efficiency, speed, and scale – elements critical to an event of such magnitude. More than just operational excellence, the system produced invaluable data for future planning, solidifying its legacy.

As a testimonial to the outstanding outcomes, a Guinness World Record related to sanitation was recorded during Kumbh Mela 2019.

Kumbh is a mammoth spiritual gathering held once every 12 years at each of four locations – Haridwar, Nashik, Ujjain, and Prayagraj. Pilgrims throng in large numbers to take a dip in the holy waters of the rivers and perform various rituals, believing that the act absolves them of all sins. In 2017, the Kumbh was inscribed in UNESCO's prestigious list of Intangible Cultural Heritage of Humanity.

Of the four locations, Prayagraj has historically attracted the largest number of visitors. In 2019, the Kumbh at Prayagraj shattered all past records, hosting an estimated 240 million pilgrims through the winter months of January and February, making it the largest ever human gathering. The population of visitors was almost 150 times the population of the host city. The principal attraction was the holy dip in the sacred waters of the Triveni Sangam, the confluence of the three sacred

DOI: 10.1201/9781003487531-26

rivers – Ganga, Yamuna, and the fabled Saraswati. On the most auspicious day, February 4, an estimated 30 million people visited.

To host this population surge, authorities constructed a tent city over 3200 hectares. The infrastructure was staggering: 440 km of temporary roads, 22 pontoon bridges, 49,968 LED street lights, 1395 km of electric wiring, and 1002 km of water pipelines.

Yet the real challenge lay in sanitation. The question was: how do you ensure an open defecation–free, odour-free, garbage-free event in the absence of precedence?

21.1 CLEANING UP KUMBH MELA

Where does one start if the target is an open defecation–free Kumbh Mela or any mass gathering of such scale? What is the infrastructure needed for sanitation? How much infrastructure should be created in a temporary setup that can to cater to an unpredictable demand? How do you ensure round-the-clock services and maintenance?

The Kumbh Melas, in the past, much like other similar mass gatherings in India, have been infamous for the poor sanitation and hygiene conditions. These have been instrumental in keeping many visitors away. Several studies have suggested that poor sanitation in mass gatherings also results in adverse impacts on health and environment.

Traditionally the toilet structures had been rudimentary and inadequate. Toilets would be located on the criteria of available space and not accessibility. Toilet servicing and maintenance remained historically poor and prone to deterioration as the mela progressed. Stench from toilets would be pervasive, a reflection of how poorly toilets were cleaned and maintained. An informal and loose system would be used to monitor sanitation infrastructure, with overemphasis on maintenance of toilet pits and not enough on services. Open defecation and manual burying of night soil was an accepted practice.

21.2 INTERVENTION ONE: SCALING UP SANITATION INFRASTRUCTURE

Magh Mela 2018 was organised in Prayagraj at the same venue as the Kumbh. It served as a crucible for many innovations in sanitation. The district administration, keen on transforming the image of the event, was persuaded to increase the number of public toilets by a factor of 10, that is, from 500 every year to 5000 in 2018. The subsequent reduction in open defecation was dramatic and evident.

Following this, the government of Uttar Pradesh decided to ramp up the number of public toilets in Kumbh 2019 to almost 80,000, a 16-fold increase from Kumbh 2013. Mela authorities also agreed to allocate space for public toilets along roadsides at frequent and regular intervals for easy visibility and access. Additionally, areas prone to open defecation were mapped and assigned for toilet installation.

21.3 INTERVENTION TWO: TRANSFORMING TOILET DESIGN

In previous melas, toilets were made either of canvas or corrugated tin sheets held together by bamboo poles. Not only was the fragile structure difficult to use, but it was also prone to crumbling as the mela progressed. Canvas or tin doors did little to guarantee privacy or security, especially for women. Light bulbs and wiring were prone to pilferage. Toilet floors, made of shoddy cement work, were difficult to clean. Sewage from toilets set up along riversides was collected in *kuccha* or brickwork pits. The risk of seepage and contamination of ground and river water was high and avoidable both from ecological and religious perspectives. It was clear that toilet designs required a transformative upgrade.

A breakthrough was made in the Magh Mela of 2018. Several modern and user-friendly toilet designs were installed and showcased. The new designs caught the attention of users, mela authorities, and the government for their obvious benefits. One year later, they became the norm for procurement of toilets during Kumbh.

21.4 INTERVENTION THREE: TACKLING ODOUR

A senior seer once remarked that walking through the mela was often a nightmare because of stench emanating from toilets and urinals. This was reason enough that, despite all campers wanting toilets, none wanted them near their camps. This was a classic case of "not in my backyard", which played a big role in the inability to locate toilets closer to users.

Unfortunately, the available solutions for odour management weren't solutions at all. Markets were flooded with options for *masking* toilet odour, none for *eliminating* it.

Magh Mela 2018 provided a fortuitous opportunity for conducting a pilot for odour management. The solution, developed by Bangalore University scientists, was tested, and its efficacy was demonstrated before local officials and other key stakeholders. One litre of this freshly prepared eco-friendly solution ensured zero stench in a public toilet for 24 hours. The solution was scaled up for Kumbh, with the help of students from Allahabad University in a pathbreaking successfully executed lab-to-field scale-up with a research–student–administration partnership. The result: near zero stench across thousands of public toilets.

21.5 INTERVENTION FOUR: ENHANCING MAINTENANCE AND SERVICES

Could the provision of adequate modern toilets alone address the problem of open defecation? Keeping the toilets clean and usable round the clock remained a critical part of the solution.

The government had procured and established almost 80,000 public toilets and urinals, of which almost 60,000 were within the mela area, while others were distributed along feeder roads and parking spaces. A quantum jump had been made in toilet numbers. However, a proportionate increase in sanitation workers was out of the question both for reasons of availability and cost. It was obvious that toilet cleaning and overall sanitation services would have to be made more efficient.

New modern toilets used in this Kumbh addressed a part of the problem, as they lent themselves to easier and more effective cleaning. In addition, jet spray machines were introduced for toilet cleaning, as they require less water and time and minimise human contact with excreta during cleaning. This were significant steps, with additional benefits of humanising work conditions.

This was supplemented by diverting scarce sanitation manpower from road sweeping to toilet services. Modern garbage management, another transformative innovation during Kumbh 2019, had considerably reduced littering. As a consequence, the burden of road sweeping came down dramatically.

21.6 INTERVENTION FIVE: SYSTEMATISING MONITORING

Our next challenge was to ensure that the toilet maintenance services worked at all times for the round-the-clock use of toilets.

How were these toilets, spread over 3200 hectares of the tent city, going to be monitored for cleanliness? Who was going to be the eyes and ears of the organising authorities? And even if a method of giving feedback was put in place, how was it going to be channelled to meet the final objective: to plug gaps in services so that toilets were clean and usable?

Traditionally, feedback on toilets or directions for corrective action were communicated during meetings or through informal channels, mostly by word of mouth. There was ample proof from past events that this mechanism was at best a partial solution. As the mela progressed, meetings and feedback become irregular, and toilet maintenance progressively deteriorated.

During Magh Mela, Prayagraj district administration agreed to depute a team of about 100 *swachhagrahis* in the mela. *Swachhagrahis* are village volunteers trained under the Rural Sanitation

Mission of the government of India to persuade village residents to adopt toilets. They are conversant with local dialects and sensitive to needs of rural citizens and thus better placed to trigger behaviour change towards toilet adoption. In the mela, their original task was to help visitors locate toilets and steer them towards using them. We experimented by assigning them the additional task of monitoring toilet maintenance. They were allotted designated areas and asked to report on the status of maintenance of toilets in their respective areas. The feedback was taken every evening between 4 to 5 PM and shared with the Health Department officials just before their daily evening departmental review meeting.

Feedback was toilet specific and on key cleanliness indicators. Each toilet was assigned a unique code for getting toilet-specific feedback, an endeavour undertaken for the first time ever. The effort proved well worth it. After the initial pushback, health officials started using the feedback to plan their tasks for the next day. Local media started publishing reports about improved toilet facilities, which reinforced the positive feedback loop.

21.7 INTERVENTION SIX: SMART TOILET MONITORING WITH IOT

Building on this experience, and taking it some notches ahead, an Internet of Things (IoT)–based smart toilet monitoring system was designed and implemented for the Kumbh Mela.

The purpose of the monitoring system was to generate feedback on toilet maintenance, which could be used for corrective action. The IoT allowed for complete transparency and reliability in two-way communication, besides being cost effective.

The first design step was the development of a toilet code format that also captured the location and service provider of the toilet. The second step was identifying monitoring indicators that were unambiguous, easy to comprehend, and adequately reflected key maintenance criteria. After much deliberation and testing, we chose eight indicators covering status of cleanliness and odour, operational status of lights and latches, and availability of cleaning staff and cleaning tools.

Third, the system was designed for two-way communication. It received feedback in real time, processed it, and sent a targeted message to the responsible ground personnel. In the actual roll out, we would start monitoring toilets as early as 4 AM. Our field staff would start receiving messages at 4 AM itself, pushing them to start their rounds to address the gaps.

Fourth and perhaps most crucially, drawing from the Magh Mela pilot of 2018, *swachhagrahis* were identified as the on-ground personnel to run the monitoring system. This required developing the monitoring app in Hindi. It also required conducting inception and mid-course training of almost 1400 *swachhagrahis*; 340 teams comprising 4 *swachhagrahis* each were assigned 1 kilometre of road and equipped with one mobile device each. Every team was required to monitor all toilets and urinals on the road assigned to them.

Last, the entire mela area, comprising 350 kilometres of roads, was divided into 340 stretches of approximately 1 kilometre each. Each stretch was assigned a unique code. This code doubled as the code of the *swachhagrahi* team assigned to that stretch and was also the code for the health department supervisor who was responsible for health and sanitation activities for that stretch. Furthermore, the codes of all public toilets on that stretch were also linked to the road code at the back end in the database. This database structure allowed us to break the entire mela area into 340 sections and assign responsibility for each section. This proved extremely effective for transparency and accountability.

Each *swachhagrahi* team would start their day at about 2 AM. They would walk to their assigned road and start the monitoring exercise by 4 AM. The first monitoring round continued until 10 AM. Once a *swachhagrahi* logged into the app using their unique code, all toilets on their road appeared on their screen. They would walk to a toilet and scan the QR code pasted on its door. The eight-question menu would drop down. Feedback, based on observations, was entered against each of the eight indicators and submitted by pressing the submit button. The feedback raced to the server,

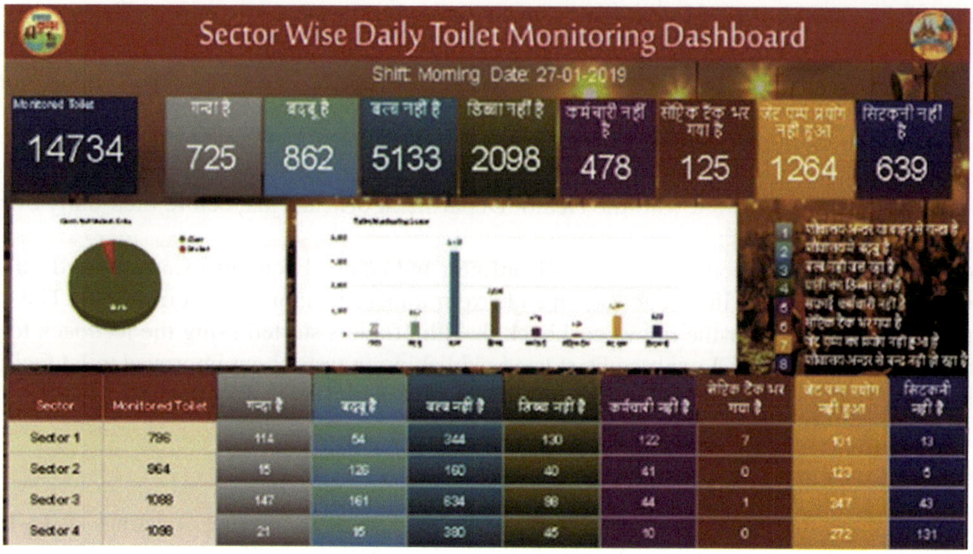

FIGURE 21.1 Sample screenshot of the sanitation monitoring dashboard.

where it got recorded. It may be noted that the server not only recorded the feedback but also the target toilet and who sent it.

Since the database had the contact information (mobile number) of the concerned health department supervisor, it automatically routed a message to that number. A typical message would read something like "X toilet has been found dirty", X being the unique toilet code. Supervisors were expected to ensure that corrective action was initiated as soon as possible.

The same process was repeated the same day in the second monitoring round, starting 2 PM and going on to 10 PM.

For illustration purposes, we present a screen shot of the dashboard in Figure 21.1.

The sanitation dashboard at any given time looked something like this. On the morning of January 27, 2019, at the time this screenshot was captured, 14,734 toilets had been monitored. Of these, 725 (less than 5%) had been found to be dirty.

This system didn't just enable monitoring of toilet maintenance services, it also had several other benefits. The health department supervisors became directly linked with the quality of services in their assigned areas. As the mela progressed and supervisors became aware that feedback was linked to their individual performance, they became more alert. It also linked the quality of reporting directly to a specific *swachhagrahi* team.

End-of-day reports, a compilation of the days' results, were shared as hard and soft copies with senior health department officials. The data and daily summary report were handy tools for all officials during the mela to acknowledge good workers and chastise the laggards. Later it also came in handy at the time of estimating financial penalties. The IoT system had generated irrefutable evidence about gaps in service delivery.

21.8 SWACHH KUMBH: FROM VISION TO REALITY

Kumbh 2019 was declared open defecation and odour free and culminated with a Guinness World Record in sanitation. As the mela progressed, reports of cleanliness published in national and international media became the biggest reason for drawing visitors, making Kumbh 2019 the most visited and inclusive event ever.

FIGURE 21.2 Receiving the Guinness World Record for Swachh Kumbh.

In the long run, the Kumbh 2019 experience offers powerful lessons for anyone shaping large-scale public initiatives. It underscores that success begins with thoughtful, purpose-driven design, and technology – when embedded early – can be a force multiplier for efficiency and accountability. But perhaps most importantly, it reminds us that the heart of every great service delivery system is human – both the one who serves and the one who is served. Keeping that human connection at the core is what transforms a project from functional to transformational.

BIBLIOGRAPHY

Andrenacci, Luciano, and Michael Barzelay. (2025). Planning and programming for a government-hosted mass-gathering event in India: the 2019 Prayagraj Kumbh Mela. In Sallai, D. & Pepper, A. (Eds.), *Navigating the 21st Century Business World: Case Studies in Management* (pp. 241–250). LSE Press. https://doi.org/10.31389/lsepress.nbw.q

Baranwal, Annu, Ankit Anand, Ravikant Singh, Mridul Deka, Abhishek Paul, Sunny Borgohain, and Nobhojit Roy. "Managing the earth's biggest mass gathering event and wash conditions: Maha Kumbh Mela (India)." *PLoS Currents* 7 (2015): ecurrents-dis.

Vortmann, Michael, Satchit Balsari, Susan R. Holman, and P. Gregg Greenough. "Water, sanitation, and hygiene at the world's largest mass gathering." *Current Infectious Disease Reports* 17, no. 2 (2015): 5.

22 Approaching Sustainable Development in Andhra Pradesh through Managing Its Water Resources

Chandrakiran Mallarapu and Nikhilesh Kumar

ABSTRACT

This chapter captures the journey of developing the Andhra Pradesh Water Resources Information and Management System (APWRIMS), which uses machine learning, artificial intelligence, and big data to depict real-time data on state water resources and present data-driven insights for decision-makers of the state. Water is managed in the state through APWRIMS, which measures the total amount of water in the state and how it spreads across the area, including major and medium projects, the purposes for usage of water, groundwater and freshwater supply and depletion levels, water pollution levels, and strategies for water conservation in the state.

22.1 INTRODUCTION

Climate change and water scarcity are problems that the entire world is facing today. From the study published in 2019 by UNESCO on water stress, it is evident that the problem is severe, as shown in Figure 22.1, which depicts a global overview of countries experiencing different levels of water stress (the ratio of total freshwater withdrawn annually by all major sectors).

While considering all these facts, extreme natural events cannot be ignored. About 90% of natural disasters are water related. From 1995–2015, floods accounted for 43% of all documented natural disasters, affecting 2.3 billion people worldwide, killing 157,000 and causing USD 662 billion in damage (CRED/UNISDR, 2015). In terms of the number of people affected, the impacts of floods, droughts, and conflicts are grossly outweighed by the number of those affected or killed by inadequate drinking water and sanitation services.

In India, the number of people affected by extreme weather, especially from the perspective of water, is immense. Around 17.5 million people were affected by drought and approximately 19 million by floods in a decade (1996–2015). Per the World Bank Group, India is among the world's most water-stressed countries. While India had 3,000–4,000 cubic meters of water per person in 1950, today it is around 1,000 cubic meters, mainly due to its population growth. The problem has intensified due to climate change, leading to uneven distribution of rainfall and ineffective water management, which may affect its food security in the future.[1]

The problem in India has more to do with water management than water availability. While the country's water resource base is substantial at 2,518 billion cubic meters, the distribution and management are skewed. For example, during the monsoon season, 50% of the annual precipitation falls in less than one month, and 90% of river flows occur in only four months, resulting in huge losses because of poor strategy and infrastructure.

DOI: 10.1201/9781003487531-27

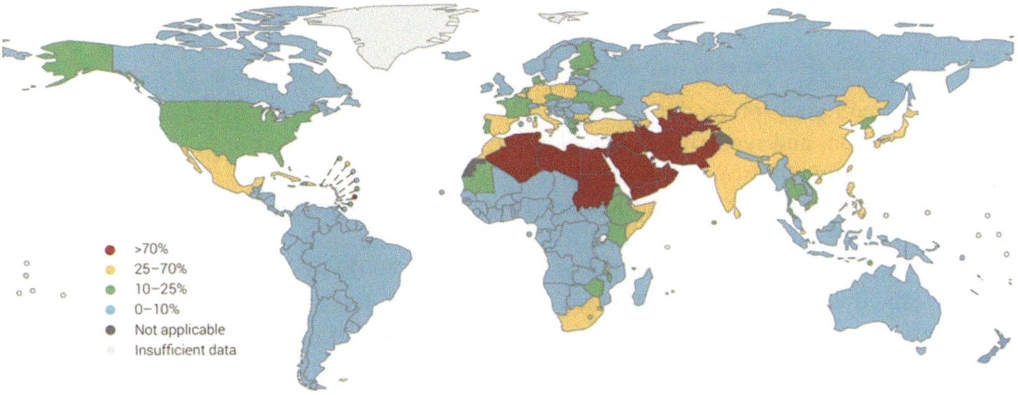

FIGURE 22.1 Water stress levels across the world (based on UN AQUASTAT, 2018).

Coupled with these problems, there is an increasing water demand, and it is expected to increase even further in the future owing to India's growing population, their demand for food, the expanding agriculture sector, the growing needs of the industrial sector, and so on.

On the policy side, recognising the impending problem and the need for long-term solutions is evident from the slew of measures initiated by the government of India (GoI) and several state governments. The process started with amalgamating several ministries, such as Water Resources, River Development, Ganga Rejuvenation, and Drinking Water and Sanitation. After amalgamation, GoI created a unified Jal Shakti Ministry under a full-fledged cabinet minister. The ministry recently initiated its flagship program, called the Jal Shakti Abhiyan, which focuses on water conservation and rainwater harvesting, renovation of traditional and other water bodies, reuse of water and recharging structures, watershed development, and intensive afforestation.

The Atal Bhujal Yojana (ABHY) is another initiative aimed at the long-term sustainability of groundwater resources in the country. The Atal Bhujal Yojana is a centrally sponsored scheme launched by the government of India to sustain groundwater resources. The scheme's primary objective is to address the critical issues of groundwater depletion and degradation in priority areas across the country. ABHY promotes community participation and decentralised groundwater management by implementing various interventions such as water budgeting, demand-side management, and groundwater recharge measures. The scheme also emphasises adopting innovative technologies and practices for efficiently utilising groundwater resources while ensuring equity and social inclusiveness. Through ABHY, the government aims to achieve the long-term sustainability of groundwater sources and enhance the resilience of communities dependent on groundwater for their livelihoods.

The NITI Aayog's Composite Water Management Index (CWMI) is an annual assessment tool launched in 2018 to evaluate the management of water resources in Indian states and union territories. The index assesses states' performance on various parameters related to water management, including water availability, infrastructure, and governance. It provides a comprehensive understanding of the water situation across the country, highlighting areas of strength and areas that require improvement. By benchmarking states against each other, the CWMI aims to foster competition and encourage states to adopt best practices for efficient water management. The index plays a crucial role in guiding policy formulation and investment decisions to address the challenges of water scarcity and promote sustainable water management practices in India. Some states have been proactive in their approach toward water management. Gujarat, Andhra Pradesh, Madhya Pradesh, and Goa were ranked highest among non-Himalayan states in the Niti Aayog's annual Composite Water Management Index report released in August 2019.[2]

22.2 CURRENT SCENARIO

22.2.1 The Case of Andhra Pradesh

The newly formed state of Andhra Pradesh (AP) faced recurring droughts, especially in the Rayalaseema region, and recurring floods along the coastal belt. Added to this is the problem of indiscriminate drilling for groundwater.

Per the government of AP's Raithu Kosam (Primary Sector Development: Status, Strategy and Action Plan), submitted to NITI Ayog in 2015, AP is one of the most affected states due to natural calamities, which have increased over the years. From 2008–09 to 2013–14, 20.18 lakh ha were affected in united Andhra Pradesh, of which 75% (15.16 lakh ha) is in new Andhra Pradesh. AP's primary sector (agriculture and allied sectors) accounts for 27% of the state GSDP and employs 46 lakh farm families, or around 62% of the state's population. So, the performance of the sector determines the progress of the state. It is also important to highlight that agriculture and allied sectors are the primary consumers of water; therefore, effective water management is imperative for the state.

22.2.2 Andhra Pradesh Water Resources Information and Management System

It is in this context that the GoAP has conceptualised an integrated approach to water management, and our team was instrumental in developing the Andhra Pradesh Water Resources Information and Management System (APWRIMS), with real-time data from various sources of water leveraging the Internet of Things (IoT), mobile apps, and satellite data. Figures 22.2, 22.3, and 22.4 show some snapshots of the APWRIMS system taken in 2020.

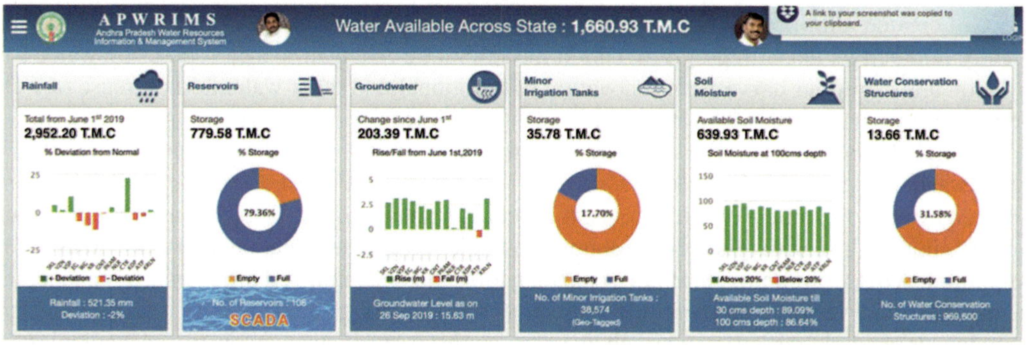

FIGURE 22.2 Snapshot of the APWRIMS (1).

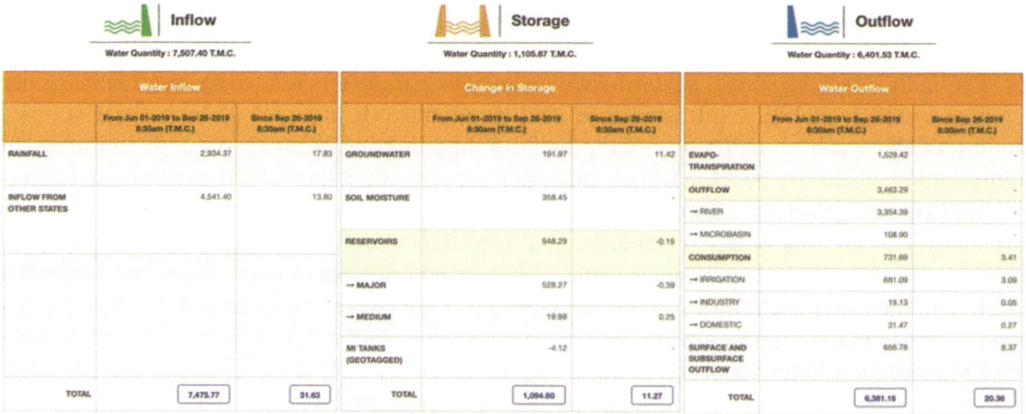

Inflow

Water Quantity : 7,507.40 T.M.C.

Water Inflow	From Jun 01-2019 to Sep 26-2019 8:30am (T.M.C.)	Since Sep 26-2019 8:30am (T.M.C.)
RAINFALL	2,934.37	17.83
INFLOW FROM OTHER STATES	4,541.40	13.80
TOTAL	7,475.77	31.63

Storage

Water Quantity : 1,105.67 T.M.C.

Change in Storage	From Jun 01-2019 to Sep 26-2019 8:30am (T.M.C.)	Since Sep 26-2019 8:30am (T.M.C.)
GROUNDWATER	191.97	11.42
SOIL MOISTURE	358.45	-
RESERVOIRS	548.29	-0.15
→ MAJOR	528.27	-0.39
→ MEDIUM	19.98	0.25
MI TANKS (GEOTAGGED)	-4.12	-
TOTAL	1,094.60	11.27

Outflow

Water Quantity : 6,401.53 T.M.C.

Water Outflow	From Jun 01-2019 to Sep 26-2019 8:30am (T.M.C.)	Since Sep 26-2019 8:30am (T.M.C.)
EVAPO-TRANSPIRATION	1,529.42	-
OUTFLOW	3,463.29	-
→ RIVER	3,354.39	-
→ MICROBASIN	108.90	-
CONSUMPTION	731.69	3.41
→ IRRIGATION	681.09	3.09
→ INDUSTRY	19.13	0.05
→ DOMESTIC	31.47	0.27
SURFACE AND SUBSURFACE OUTFLOW	656.78	8.37
TOTAL	6,381.18	20.36

FIGURE 22.3 Snapshot of the APWRIMS (2).

FIGURE 22.4 Snapshot of the APWRIMS (3).

The solution includes reservoir water storages of major and medium projects and minor projects, minor tanks, water conservation structures (check dams, percolation tanks), groundwater, and most importantly soil moisture. The dashboards depict real-time data, along with data-driven insights. The diverse data sets are collated and analysed using tools like machine learning, artificial intelligence, and big data.

The key insights provided by the solution are:

- How much water do we have now?
- How is it spread across reservoirs, groundwater, soil moisture, minor irrigation tanks, and so on?
- How much water is used and for what purposes across the year?
- What are the groundwater and freshwater usage and depletion levels?
- Which villages are running under water deficits?
- Which locations are affected by drought?
- What can be done to conserve water?
- What are the right locations to put conservation structures?

In this case, these key insights were provided for the state of Andhra Pradesh. They are generic, however, and can be replicated for any state facing similar water-related issues.

Further, the relevance of a solution to view and manage water resources on a real-time basis is due to the following reasons:

- Freshwater is limited and essential for sustainable growth.
- Water resource management should be based on users, planners, and policymakers' participation.
- Water has an economic value, and it's a fundamental human right to access clean water.
- The human and economic impacts of flood and drought periods can be particularly high, and information systems can help mitigate the risk and reduce these impacts.

Managing water resources in transboundary basins requires sharing data and information that meets stakeholders' expectations for activities like planning, monitoring, assessment, and prevention alerts. Unfortunately, the exchange of information and data on a transboundary basin is often tricky both for structural reasons (when there is no agreement or protocol between the different authorities) and for technical reasons (difficulties related to information collection, harmonisation, and data processing), as well as standard problems (dispersed, heterogeneous, incomplete and rarely comparable data). In our union of states structure, often state authorities are reluctant to provide neighbouring states with information they consider strategic. In addition, the economic value of water used for hydropower, agricultural irrigation, and navigation may increase this reluctance.

22.2.3 Drawbacks of Current Water Management

First, it must be highlighted that until now, the implementation of decisions like canal water releases and construction of conservation structures are based on historical data. Further, using expert inputs, the estimation of unaccounted water resources is done by applying a multiplication factor. The data and insights are manually correlated from different systems, and water management decisions are based on these.

Another specific issue is that the estimation of the overall runoff is sometimes done without considering the water conservation structures upstream. Then there is poor participation from all the stakeholders in managing water resources. Often outdated data is used to estimate the current outlook. The local administrative decision-making doesn't consider basin-level management. It can be said that there is more projection rather than a prediction of disaster events like drought and floods based on historical data.

Overall, this reduces the scope for automation to optimise repetitive steps of the process. To conclude, there was no data-driven decision making (non-scientific approaches in many places), which riddled the entire process with errors and assumptions.

With multiple levels of authority involved in water management in the state, it is imperative to have a unified solution for visibility to make quick and informed decisions. As a reference, here is the indicative structure of the water management departments.

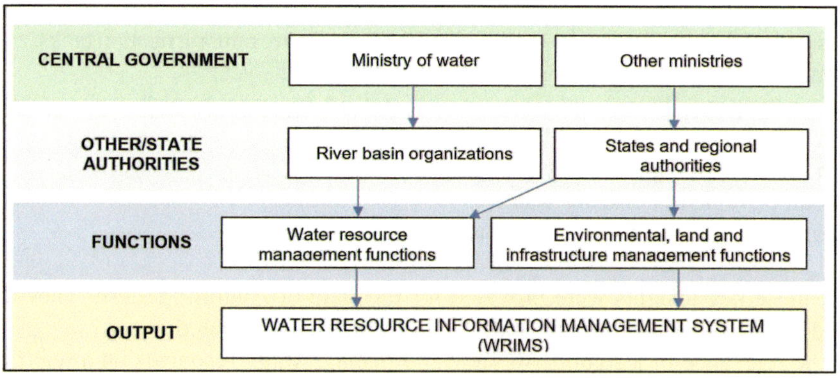

FIGURE 22.5 Structure of water management departments.

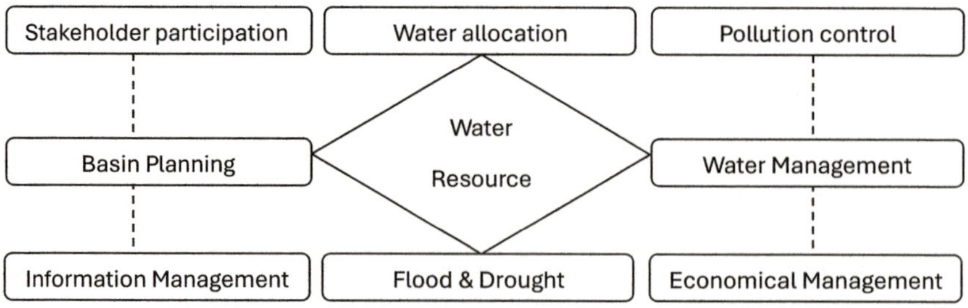

FIGURE 22.6 Facets of water resource management.

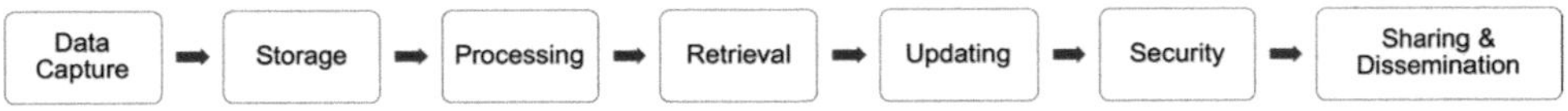

FIGURE 22.7 Workflow of the information management system.

Comprehensive information management systems are, thus, key instruments for developing integrated management of transboundary basins since they are tools designed to facilitate the production and sharing of information expected by stakeholders.

With consideration of this flow, the Water Resource Information Management System (WRIMS) aims at synchronising all units under one umbrella to collect data from every source and return meaningful insights for all stakeholders, as illustrated in the following.

To achieve this, it requires work towards getting and understanding the data and applying artificial intelligence and machine learning algorithms to automate the process and eliminate constant human intervention.

22.2.4 Methodology, Data, and Technology

An effective WRIMS needs an objective basis for planning to gain the confidence of all the stakeholders and the public, in which information has to be organised, retrieved, acquired, and maintained or updated. To achieve the desired results, a very basic workflow of information management (Figure 22.7) will have to be ensured, which may become complicated on account of complexities over time.

While developing the Andhra Pradesh WRIMS, during the data capture process, our team collected information from approximately 1 million geo-tagged water bodies and 1.3 million geo-tagged bore wells, along with various layers like administrative, hydrological, soil, agriculture, satellite, aquifer, land use land cover (LULC), rivers and canals, drainage networks, digital elevation model (DEM) data, and command areas.

22.2.5 Challenges

Approximately 3 lakh data points come every day from different sources like automatic weather stations (AWSs), reservoir-level sensors, ground water–level sensors, and river and canal–level sensors to the APWRIMS. Information for different data points comes from satellites, models, the India Meteorological Department (IMD), Indian Space Research Organization (ISRO), and Water Resources Division (WRD), among other sources. All this data arrives in different formats and frequencies and, hence, poses a big challenge to sanitise, validate, and feed into the system for actual impact.

The following figure 22.9 illustrates the data required to build a comprehensive solution like APWRIMS.

For states with data-poor environments and a lack of sensor networks, a basic implementation of WRIMS needs to be and can be built with satellite data and other open data sources. All the components in the APWRIMS are built on open source components, which helps reduce the system's development cost. Big data technologies are used to ensure the system can scale horizontally and handle large volumes of data. A Cassandra big data database (a big data database used to store large volume of data per project requirements) was set up as a three-node active cluster (which can be expanded to include more nodes as the data expands) so that there is no single point of failure and the data can be available in high availability architecture.

All time-series data would be stored in big data database, whereas the configuration data will be stored in a relational database. Both the big data database as well as a relational database used are open-source databases. Other development tools used are JAVA, R, Python, Angular JS, Node JS, HTML, and CSS. The platform is built with future expandability designed into the architecture that will allow easy vertical and horizontal scaling. The system is deployed with a low cost of operation and maintenance structure.

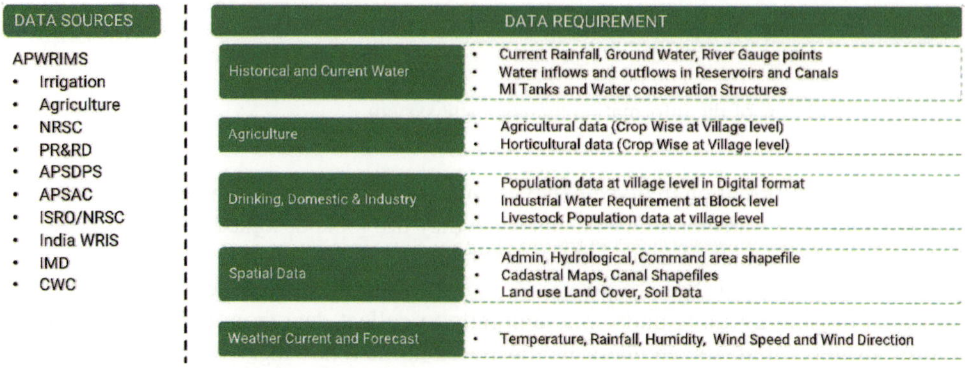

FIGURE 22.8 Data required to build APWRIMS.

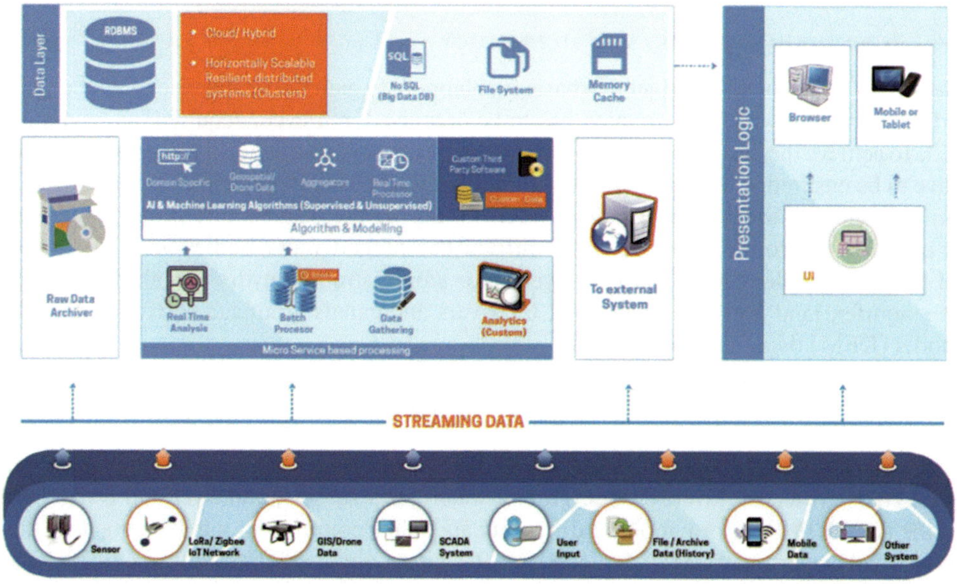

FIGURE 22.9 Database and data flows of the system.

22.2.6 Types of Data and Understanding the Data

The data is collected and processed to derive three benefits – water resource information, water resource management, and geoportal views. The objective of this stage is to produce results and dashboards made for decision making and empowering stakeholders. Figure 22.10 shows the data collection and processing for decision support.

To meet the stated objective, collecting the data was the first step. The data had a lot of inconsistencies as well as missing records. APWRIMS includes algorithms and models to clean the data and retrieve and augment the data. Further, it integrates data from public sources, such as satellite data augmented with machine learning and hydrology/crop-specific mathematical models, to fill any gaps. Various software modules then translate the data to easily readable dashboards with GIS views and provide decision support on top of it. There are three primary types of data that are used in various analyses. These are collected through multiple sources.

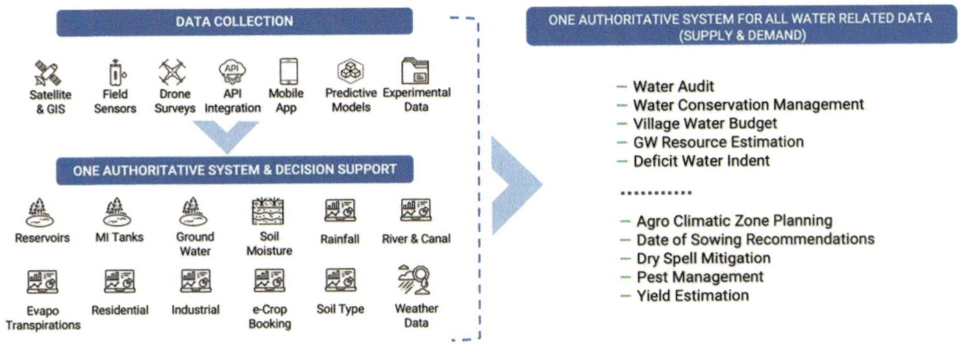

FIGURE 22.10 Data collection and processing.

- **Master Data:** The system consists of various static information from spatial and non-spatial perspectives. These can be geo-tagged water bodies or geotagged bore wells, along with various layers like administrative, hydrological, soil, agriculture, satellite, aquifer, LULC, rivers and canals, drainage networks, digital elevation models, and command areas.
- **Dynamic Data:** The dynamic data are from various sources such as sensors, mobile apps, third-party web services/web sites, and satellite data. The data frequency of these data sources can be as high as every hour or more frequent in the times of floods.
- **Historical Data:** Historical data like weather, surface water body levels, groundwater levels, and agricultural data in various formats are all harmonised, classified, and stored in system databases for analysis and usage in predictive models.

The wide-ranging geospatial data organised in WRIMS comprises both primary and secondary data, containing:

- **Base Map**: Open street, Bhuvan, DEM, Stamen terrain, satellite images
- **Thematic**: NDVI, soil moisture, NDWI, contours, LULC
- **Hydromet Spatial Data**: Rainfall, groundwater, soil moisture
- **Surface Water**: Reservoirs, dams, barrages, rivers, streams, canals, minor irrigation tanks
- **Water Conservation structures**: Check dams, percolation tanks, form ponds

The power of the IoT is leveraged by getting data from following types of sensors in near real time:

- Piezometers with digital water level recorders and telemetry for groundwater levels monitoring on a near real-time basis (about 1,250 sensors across the state of Andhra Pradesh)
- Automated weather station sensors for monitoring weather parameters including rainfall on a real-time basis (about 1900 stations across Andhra Pradesh)
- Reservoir-level sensors for monitoring near real-time levels and storages for 100+ reservoirs
- About 900+ soil moisture sensors monitoring soil moisture at three different depths on a real-time basis
- Water quality monitoring
- Flow sensors using Doppler sensors for measuring flow – especially in some critical lift irrigation schemes

The collected data is used in models to build the necessary decision support. The models used in APWRIMS were created with the help of domain experts from organisations such as the National

Remote Sensing Centre (NRSC), International Crops Research Institute for the Semi-Arid Tropics (ICRISAT), Acharya N. G. Ranga Agricultural University, Andhra Pradesh (ANGRAU), and various other line departments.

APWRIMS uses many models to process the data for specific purposes. Some examples are shown in Table 22.1.

Data is used as a first step to build visibility dashboards in both management information system (MIS) and GIS views for rainfall, reservoirs, tanks, groundwater, and soil moisture. The geoportal consists of an integrated, user-friendly dashboard that enables high-level officials to look at the GIS map view of various layers from admin boundaries and political, agricultural, demographic, and thematic layers.

The design incorporates visualisation techniques such as circle views, grid layout, small multiple maps, and node simplification to improve the data readability of water distribution systems from the macro to micro level, with an intuitive geoportal. The geoportal serves the purpose of one authoritative system for all water-related data supply, demand, operational, and environmental factors with near real-time visibility into 90% of all available water such as groundwater and surface water components such as reservoirs, canals, and minor irrigation tanks.

Figure 22.11 shows the technical stack of APWRIMS.

TABLE 22.1
Applications of Various Hydrology Models

Hydrology Model	Use of Model
Variable Infiltration Capacity Model (VIC)	Rainfall to runoff value calculations
VIC Routing Model (RVIC)	To create drain network and flow accumulation
Network Flow Model	To calculate runoff values from last 30-yr rainfall data
Crop-Specific Water Balance Model	To calculate crop-specific water demands
GEC Model	To estimate groundwater
Penman Monteith	To determine reference evapotranspiration
Various optimisation algorithms	Agro eco-crop planning models

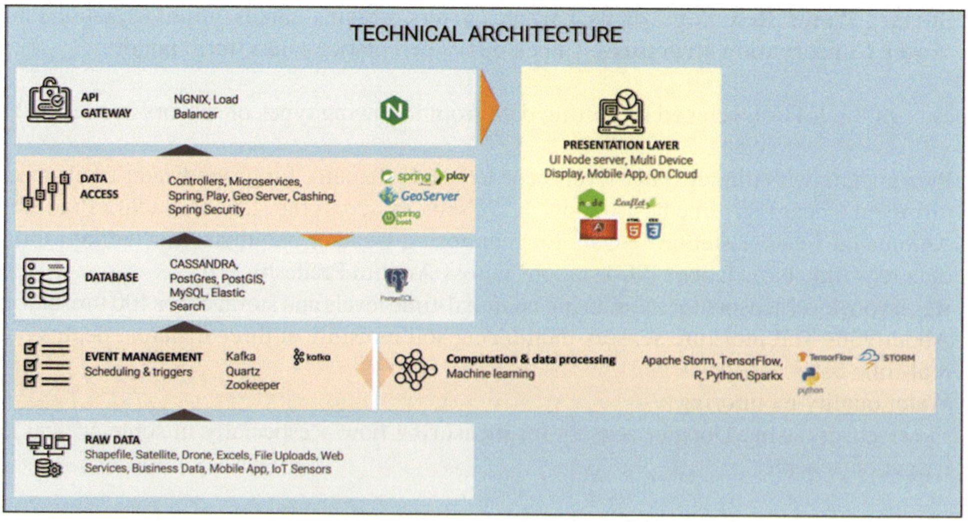

FIGURE 22.11 Technical stack of APWRIMS.

22.2.7 Decision Support System (Web Portal)

The system must be web enabled and cloud hosted, which makes it accessible from anywhere. The APWRIMS system comprises various components that get data from various sources and is automated in such a way that no manual intervention is required to disseminate outputs. This ensures seamless automation of inputs and outputs, making the system self-sustainable.

The WRIMS has to be created for all the stakeholders of the water and agricultural sectors. These include users at all levels such as planners, key decision makers, researchers, farmers, and the common public. There are user-specific dashboards, advisories, and information available. The user interface is done in vernacular GIS and easy MIS formats for the user to comprehend.

An interactive decision support system (DSS) can help experts prepare water resource management plans for decision makers and stakeholders. The structure of our DSS can be understood from the block diagram given in Figure 22.12.

The following are the key components for the APWRIMS geoportal.

22.2.8 Water-Related Components

22.2.8.1 Water Balance and Water Audit

A water audit provides a data-driven hydrological framework that facilitates traceability and accounting of the water resources across the state. It is a realistic assessment of supply, storage, usages, losses, and outflows in the present or in an earlier "water year". This module results in holistic water auditing and accounting that can help authorities for better management of water resources and planning long-term project-based interventions.

22.2.8.2 Village Water Budgets

An interactive decision support system at a village level is based on current supply and demand to help give visibility over the surplus and deficit of water at the village level.

22.2.8.3 Groundwater Resource Estimation

Estimating groundwater resources was done manually before groundwater assessment, categorisation, and evaluation (GRACE) was introduced, and there is a need to automate it. This module helps in understanding and visualising the areas that are at various stages of exploitation of groundwater.

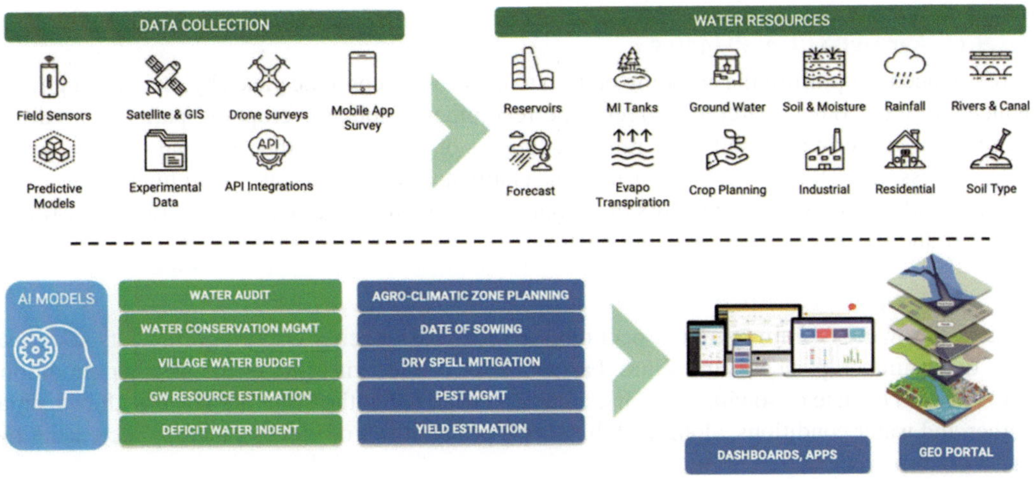

FIGURE 22.12 Structure of decision support system.

This module results in identifying the villages and areas where dynamic GW is overexploited, underexploited, and safe. Also, better water conservation management based on GW recharge, draft, and potential would be helpful.

22.2.8.4 Minor Irrigation Tank Vegetation Estimation

There is a need to find the infiltration and the extent of infiltration in and around minor irrigation tanks. Key stakeholders get information regarding the extent of vegetation in minor irrigation tanks, thus impacting the water storage capacity. This module results in decision support for corrective actions to be planned and taken to ensure that minor irrigation tanks are used to their maximum potential.

22.2.8.5 Reservoir Management

The predictive decision support system has all the reservoir related information such as estimation of inflows and outflows based on rainfall-runoff, demand, gross capacity, current storage and flood cushion for each reservoir type and river basin. Based on predicted and current inflows, demand, flood cushion, and other parameters, reservoir releases are optimised. This module results in a unified dashboard that gives a holistic view of the reservoir, enabling optimised management of reservoirs. Machine learning algorithms are used for predicting inflows at different critical points.

22.2.8.6 Soil Moisture–Based Irrigation and Canal Management

APWRIMS provides visibility of the water flow at important points in the irrigation canal network and provides decision support to advise releases based on soil moisture and crop water requirements. This module results in transparency in water flows at different critical points in the canal network and advisories based on soil moisture and required water.

22.2.8.7 Watershed Management

There is a need for better management of watersheds where there is a decision support system that looks at rainfall-runoff, slope, LULC, and other parameters and provides decision support on best possible sites for a water conservation structure and effectiveness of existing structures. This module results in identification of new water conservation structures, a system to evaluate the effectiveness of existing water conservation structures.

22.2.9 Agriculture-Related Components

22.2.9.1 Agro-Eco Crop Planning

There is a need to optimise the cropping pattern to maximise economic yields by maintaining water balance, soil, and other scarce resources. The recommendations for cropping patterns would be given after considering the soil type, rainfall, sources of water, and other external factors such as market access, market linkage, after-farm value chain, access to seeds, access to finance, access to and field workers. This module results in cropping pattern advisories based on various constraints like water and soil to optimise economic yield.

22.2.9.2 Crop Date of Sowing Advisory

The date of sowing is an important aspect of crop yields, as the crop water requirements at critical stages determine crop success or failure. Farmers have limited visibility of the weather conditions in the future. The date of sowing advisories module ensures that the crops are sown with congenial weather and water conditions, along with historical probabilities of success, thereby increasing the chance of crop success.

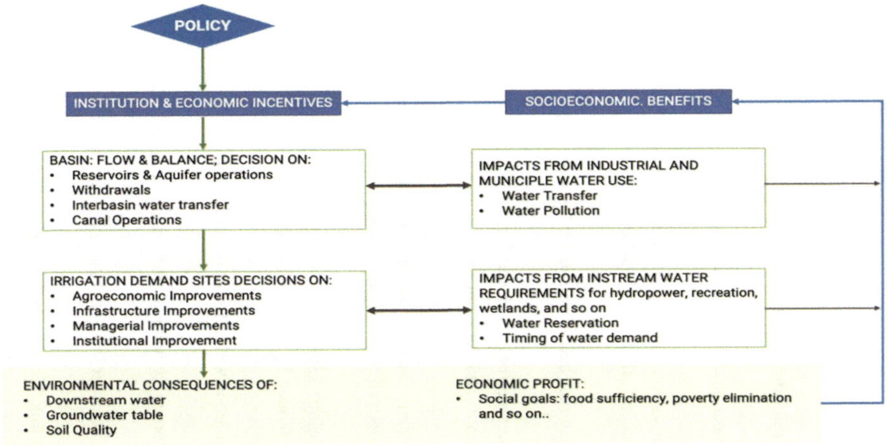

FIGURE 22.13 Salient features of WRIMS.

22.2.9.3 Crop Soil Moisture Stress Advisories

Soil moisture stress affects crop yields to a large extent. Timely interventions to mitigate soil moisture stress provide both protective and productive irrigation. This module results in timely advisories for mitigating crop stress and advisories to intervene.

Salient features of WRIMS for policy making are illustrated in Figure 22.13, which shows how data can empower decisions at a much larger scale.

22.3 IMPACT

To understand the decisions and its impact, first, we need to understand the different functions, their objectives, and their indicators. As part of a program of capacity building, support indicators have been developed that are based on the implementation of the integrated approach to the sustainable management of water resources. The indicators are presented as a minimum set and therefore do not completely measure the objectives described for good water resource management.

For example, the state objective in terms of water allocation is to manage major water users in the state, allocate water based on equality, and promote its sustainable use. For this purpose, all groundwater resources were measured so that progress can be measured in the future. The limits of water extraction were set, and exploiting water resources was clearly defined so that the state could take necessary action. Reservoirs were monitored, and the allocation of water to subunits was also reviewed from time to time.

Another facet signifying the impact of the solution is when it comes to monitoring pollution levels and the effectiveness of allocation. The solution created its own metrics in terms of recording the total number of monitoring visits. Out of these, the number where standards imposed were not being complied with were also recorded. Finally, the overall monitoring unit was arrived at by dividing the number of non-complying visits by the total number of visits.

22.3.1 Specific Impacts of APWRIMS

Overall, APWRIMS is expected to benefit more than 60% of the state's population, who are agriculturally dependent. The crop planning activity recommended was to shift from water-thirsty

TABLE 22.2
Functions and Evaluation Metrics of the Water Management System

Function	Objective	Progress Indicator	Unit/Definition
Water allocation	Management of major water users like authorities	1. View of all groundwater resources	Review Extraction limits and exploitation status
	Water allocation for sustainable use, economic efficiency, and on equality principles	2. Allocation criteria based on demand	Review Examine allocation criteria, for example, soil moisture stress
		3. Social reserves	Review Monitoring reservoirs and understanding flow and need for conservation and allocations to subunits
Pollution control	Extent of water pollution	4. Percent of surface water quality	% Number of samples below set standard
		5. Percent of groundwater quality	% Number of samples below set standard
	Major polluters and pollutants	6. Type of pollution	List of pollutants and their causes
Monitoring	Effectiveness of allocation system	7. Proportion of water allocation	% From monitoring visits, the number not complying with conditions divided by the total number of visits
	Pollution control	8. Pollution levels	% From monitoring visits, the number not complying with conditions divided by the total number of visits
	Knowledge of water resource availability	9. Number of water resource monitoring status	Number Number of stations with reliable data records
		10. Total water storage capacity	mm^3 The water storage capacity in artificial storage structures
		11. Groundwater stations reporting negative balance	% Comparison of water levels over a five-year period
Basin planning	Basin planning synthesises technical and social priorities for the basin and acts as a basis for action and accountability to stakeholders	12. Water management activities driven by basin plan	Review Examine the link between the basin plan and current water management activities
		13. Stakeholder priorities reflected in the basin plan	Review Examine the basin plan for stakeholder consultation and content

Information management	Essential data is processed and packaged as information at the right level for specific managers and stakeholders to support transparent decision making and to gain commitment and political support for the decisions made	14. Database is established in formats compatible with other river basin organisations	Review Database is transferable across basins in the country and for transboundary systems
		15. Water management information is available to managers and other stakeholders as required	Review Examine the availability of basin information and reports on water resource management indicators
Stakeholder participation	Effective cooperation between government agencies with responsibilities for water management or water use in the basin	16. Synchronising different units inside government	Review Data inflow and frequency of information gathered from various departments
	Stakeholder participation is institutionalised in the management of the river basin	17. Clear roles and responsibilities in water resources management	Review. Examine basin water management structure for stakeholder organisations and allocated management roles
		18. Representation in decision making bodies at all levels	Review Participations from NGO, water societies, public reps, and authorities

agriculture crops to suitable horticulture crops. This resulted in an increase of about 1.85 L ha of horticulture crops. The groundwater levels also improved by 2 metres across the state despite receiving a 14% deficit of rainfall.

The APWRIMS has helped to optimise the inter-basin transfer of water that provides critical and necessary water to the entire Krishna Delta region, impacting 1.1 million acres. Also, 100k+ hectares have been reported for critical soil moisture stress, and due to intervention, about 4,540 farmers benefitted.

This initiative has also helped groundnut farmers to increase their yield by 23%. It has saved 970 MW hours of energy for pumping groundwater for irrigation purposes, which costs about INR 4,850 million.

Further, the APWRIMS has helped stabilise an ayacut (area served by an irrigation process, like canals, dams, etc.) of 7.11 lakh acres.

Regarding lift scheme management: After the revival of the LI schemes, the ayacut irrigated increased substantially from 3.81 to 6.15 lakh acres. Finally, based on the decision support system, we raised advisories that would lead to a total water saving potential of 54.3 T.M.C. at 40% reliability of excess runoff.

The APWRIMS has tremendously impacted water resources and all connected sectors of life in Andhra Pradesh.

NOTES

1 https://www.worldbank.org/en/news/feature/2019/03/22/helping-india-manage-its-complex-water-resources

2 https://www.hindustantimes.com/india-news/gujarat-andhra-pradesh-top-niti-aayog-s-water-index-report/story-2JDfK5WgRw5LXga8VF6sGI.html

REFERENCES

1. Andhra Pradesh Water Resource Information Management System. www.apwrims.com
2. http://www.pacificwater.org/userfiles/file/IWRM/Toolboxes/introduction%20to%20iwrm/Tutorial_text.pdf
3. https://www.iucn.org/downloads/iwrm_a_new_way_forward_1.pdf
4. https://unesdoc.unesco.org/ark:/48223/pf0000367306
5. Johnson, Lynn E. "Water resource management decision support systems." *Journal of Water Resources Planning and Management* 112, no. 3 (1986): 308–325.
6. Pierluigi Cau, Simone Manca, Costantino Soru, Davide Muroni, Dorian Gorgan, Victor Bacu, Anthony Lehman, Nicolas Ray. and Gregory Giuliani. "An Interoperable, GIS-oriented, Information and Support System for Water Resources Management" *International Journal of Advanced Computer Science and Applications (IJACSA), EnviroGRIDS* Special Issue on "Building a Regional Observation System in the Black Sea Catchment", 2013. http://dx.doi.org/10.14569/SpecialIssue.2013.030309
7. Stefanovic, Nenad, Ivana Radojevic, Aleksandar Ostojic, Ljiljana Comic, and Marina Topuzovic. "Composite web information system for management of water resources." *Water Resources Management* 29, no. 7 (2015): 2285–2301.
8. Tsihrintzis, Vassilios A., Rizwan Hamid, and Hector R. Fuentes. "Use of geographic information systems (GIS) in water resources: A review." *Water Resources Management* 10, no. 4 (1996): 251–277.

Section V

Smart City and Urban Development

23 Using Drones to Map and Assess Road Quality in Rural Andhra Pradesh

Jayavaradhan Sambedu

ABSTRACT

Road quality assessment is a time-consuming and labour-intensive activity that requires surveying the road and capturing observations on paper and then reporting them on a cartographic map. Curl Analytics adopted an innovative pilot approach in collaboration with the government of Andhra Pradesh, where drones were flown over the road to capture images of the road, and the depth of the potholes was measured using the onboarded LiDAR. Machine learning methods were used to classify road segments into good, average, bad, or mud roads, along with information on the line markings. This information, along with the dimensions of potholes, is reported in digital format to the state authorities for real-time monitoring of road conditions and taking up repair activities.

23.1 INTRODUCTION

Roadways are a crucial component in the transportation network. For a growing economy, the construction of new roads, as well as maintaining older ones, is a continuous process. Asphalt roads (composed of bitumen, gravel, etc.), deteriorate over time, owing to temperature, oxidation, and water logging and from heavy loads that pass over them. Therefore, regular road surveys are needed to monitor road conditions.

23.2 CURRENT SCENARIO

A conventional road survey involves topographic surveys by professional land surveyors using either digital global positioning systems (DGPSs) or the total station method. These methods include physical verification and marking the roads. Given that these methods are mostly physical, the number of roads that can be surveyed each day is limited by these methods. On completion of the survey, detailed paper reports are prepared and forwarded to the concerned ministry, which takes decisions on the next actions with respect to the roads surveyed. The entire process is labour intensive and time consuming and hence ends up being done only for main highways, leaving out secondary highways, which end up being rarely inspected. However, all roads need periodic surveys.

23.3 SOLUTION

In order to overcome these shortcomings, we propose a method of imaging and precise distance measuring techniques that connects in real time to the real-time governance (RTG) system of the government of Andhra Pradesh. The solution, deployed using drones, can speed up the decision process from days in conventional methods to a matter of hours. The use of geo-referenced digital techniques removes ambiguity with respect to location information while providing speed and accuracy.

DOI: 10.1201/9781003487531-29

Thus, surveys of minor roads that were performed once every few years could now be performed once every quarter. Information on periodic quantitative and qualitative assessment of the roads, linked to the contractors who laid the roads, can be used to rate the contractors and used as one of the evaluation criteria for assigning future projects.

23.3.1 Drones in Survey (Road Quality Mapping and Assessment)

Drones are being used for a variety of use cases such as security, land encroachment detection, law and order, policing, agriculture, cargo delivery, imaging, and a plethora of other use cases.

For road monitoring, the usage of drones, equipped with a camera, Light Detection and Ranging (LiDAR), global navigation satellite system (GNSS, and edge computing infrastructure) can vastly improve road quality mapping and assessment and would go a long way in delivering real-time inputs to the Real Time Governance (RTG) system. These inputs can be used by policymakers to make decisions pertaining to budgets and priority approval of projects.

FIGURE 23.1 UAV flying.

FIGURE 23.2 NISA stationary on ground.

There are many benefits of using unmanned aerial vehicles (UAVs) or drones for such interventions. One, such drones can vastly reduce the time spent on projects by collecting accurate surveying data in a much shorter time frame, reducing the ROI significantly. Two, drones also eliminate the need for humans to access hard-to-reach and dangerous terrain physically. Simply put, unmanned aircraft system (UAS) technology makes the entire process scalable and efficient.

For the specific intervention, we employ Nisa, a custom-built drone by Curl Analytics, for the purpose of road quality assessment. This drone is enabled with a high-definition (HD) camera, light detection and ranging (LIDAR) system and global navigation satellite system sensors to capture precision latitude and longitude information. On roads to be surveyed, the three sensor systems onboard the drone acquire dense 3D and 2D data in a single flight.

We have used an edge computing enabling processor, which hosts machine learning and deep learning models developed by Curl Analytics. These models can identify potholes with precise geo-referenced information. These models can also determine the dimensions and depth of the potholes. All these inputs are then used to gauge the material required to fill these potholes. This information, with additional analytics (such as impact on budgets), is then provided to the RTG to the government of Andhra Pradesh.

The paper trail and the corresponding delay, as seen in the conventional method, are eliminated, and the key stakeholders can decide roadworks, which can be promptly conveyed to the concerned engineers/contractors at the site.

23.3.2 Nisa: Process Flow

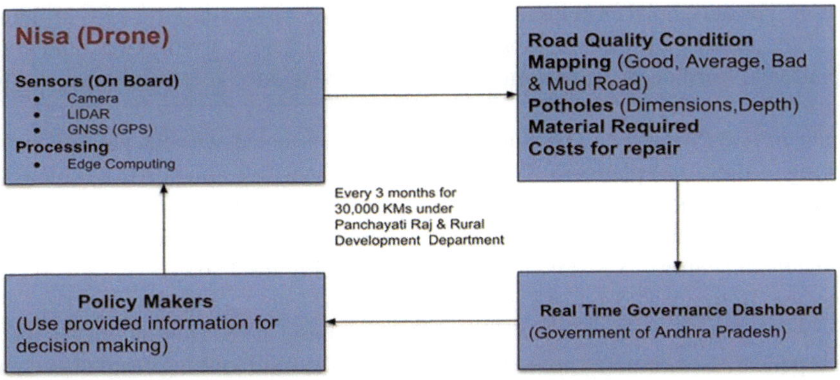

FIGURE 23.3 NISA drone process flow.

23.4 NISA FLOW MAP: DESIGN

The components for developing Nisa are as follows.

Aerial System: A system capable of performing autonomous take-off, performing the survey, and returning to the designated location adhering to the safety constraints while being robust to winds, gusts, and varying environments.

Power Management System: A system to supply regulated and continuous power to the aerial systems, and has facility for redundancies and auxiliary power supplies for crucial flight management systems, ensuring power supply to the other systems as needed. The power management system, while consuming minimum power, should ensure stable hover and be free from interference from other power sources such as high power lines it encounters during flight.

Sensors System: The sensors on board include LiDAR, an HD camera, and a GNSS system. The sensors are powered by the onboard power management system. A microprocessor onboard powers up, synchronises the different sensors, and stores the recorded data.

Weight Build-up Chart: The complete payload onboard can be classified into one of three classes – electronic support system weight, aerial frame weight, and sensor subsystem weight. The placement of the payload components on Nisa is done keeping two goals in mind: ensure equal distribution of weight along the aerial frame for stability during flight and that complementary system components are close to each other, ensuring minimum wire length and minimum delay. The all up weight (AUW) of the aerial system is determined by combining the weights of the individual subsystems. The AUW with a margin forms the basis for the design of the propulsion system.

Propulsion System Design: Design of the propulsion system involves selection of motors, propellers, battery, and design of the power distribution board.

Mechanical Design: The design of the drone is intuitive and ensures secure payload attachment. The LiDAR must be mounted facing downwards. A variety of frame types were shortlisted and analysed. Based on accessibility of the payload area on the fuselage, an H-frame was chosen, and sizing of the drone was completed. The weight build-up chart was used to estimate the sizing of the drone. A 3D-CAED model of the design was then done to verify the conceptual design to scale. Material selection was a crucial part of this design phase, as the weight of the frame must be optimised. The design was an interlocking setup with fixtures and carbon fibre tubes holding the frame together. An acrylic pre-build prototype was manufactured to ensure the assembly issue were identified, and this enabled the design to be validated and finalised.

Initial Prototype: The initial prototype was built with pre-impregnated carbon-fibre sheets CNC-cut using an industrial CNC router. This was outsourced to a professional routing firm. The assembly was done at Curl Analytics labs. The electronics on board were assembled, wired, and routed.

Calibration: The aerial platform has a F4 flight controller which supports a PX4 flight stack. The calibration of the aerial platform was done for the accelerometers, gyroscopes, and compasses of the aerial platform. Further, a calibration of the LiDAR was done to ensure the LiDAR data was optimally streamed.

Testing the Initial Prototype: The initial prototype was tested without payload just to ensure proper working of the aerial platform. The LiDAR, being sensitive and costly equipment, was then installed on the drone, and the data stream was recorded for further analysis of point cloud data.

Following are some photographs of Nisa being assembled by a team member (Figure 23.4), Nisa flying over the team (Figure 23.5), and a close-up of Nisa (Figure 23.6).

Mounting the Sensors: The LiDAR should have a 360° view, with no obstructions, between itself and the road perpendicularly below. Hence, it is mounted at the front of Nisa at the plane area below the propellers. The camera should be able to view the road looking from above and hence is mounted in the underbelly of the drone.

FIGURE 23.4 Team member assembling NISA drone.

FIGURE 23.5 Trial Runs of NISA drone.

23.4.1 Testing On-Site and Dry Run

Following are some photographs of Nisa demonstrating its various data gathering capabilities.

23.4.2 Nisa Flowmap: In Action

FIGURE 23.6 Nisa flying on road for dry run.

FIGURE 23.7 Nisa data captured from LiDAR (a).

FIGURE 23.8 Nisa data captured from LiDAR (b).

FIGURE 23.9 Snapshot of road captured by Nisa.

FIGURE 23.10 Nisa data with GNSS (position information).

23.4.3 NISA: DATA ANALYSIS: HOW IS IT DONE?

First, the camera, LiDAR, and GNSS data are synchronised. The 2D video data from the camera is used for segmentation of the road to classify it according to the quality of the road. The 3D point cloud data from LiDAR is used to determine potholes on the road. For the determined potholes, depth, extent of the pothole, material required, and cost of repair are calculated. The GPS data from GNSS is used to pinpoint the location of the survey, overlay it on a digital map, and render it on the RTG system.

FIGURE 23.11 Raw data from camera.

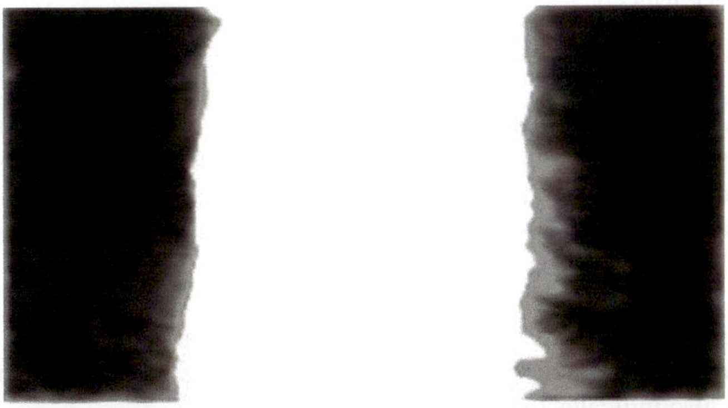

FIGURE 23.12 Deep learning based segmentation.

FIGURE 23.13 Road outline pre-processed.

FIGURE 23.14 LiDAR data (1).

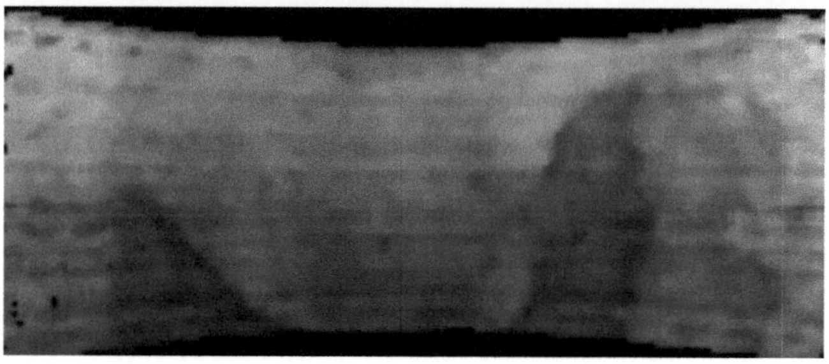

FIGURE 23.15 LiDAR data (2).

Camera Data Analysis: The video feed acquired by the camera can contain terrain features beyond the road, as seen from the picture. We used a deep learning model to classify the video feed into road regions and other terrain regions. The road region is then segmented, and a deep learning model is used to classify the road into one of four classes – good road (GR), average road (AR), mud road (MR), and bad road (BR). In addition, it is trained to identify lane markings on the roads.

Following are some photographs of camera data captured from a single road.

In the given analysis, the model provides a confidence score of 99.999% that it is viewing an average road, and it also indicates that there are *no* lane markings.

LiDAR Data Analysis: LiDAR produces a dense 3D point cloud of relative GNSS locations, as shown previously. To use the information provided, we convert this 3D point cloud to a 2D intensity map that is used for computation. The following images provide us insights on how LiDAR intensity varies as it flies over a pothole. The discoloration indicates intensity, which is mathematically related to depth information.

23.5 IMPACT

Nisa's pilot run was performed on Hindupur Road (Anantpur district). Nisa was flown post all the calibrations that were performed on its systems. First, the raw data captured from the onboard HD camera was processed to segment the road. From the 3D point cloud generated from LiDAR, we converted a 2D intensity map, and the GNSS information from the drone was overlaid to identify the precise location of the potholes. This information was then plotted on a map along with the estimation of material required to fill up the pothole and the costs involved in road maintenance. Some details of the intervention are provided in the following figure.

District: Anantapur; location: Hindupur Road

FIGURE 23.16 GNSS co-ordinates (13.9630652 N, 77.64066 E to 13.9663203 N, 77.6400389 E).

BIBLIOGRAPHY

Askarzadeh, Taraneh, Raj Bridgelall, and Denver D. Tolliver. "Drones for road condition monitoring: Applications and benefits." *Journal of Transportation Engineering, Part B: Pavements* 151, no. 1 (2025): 04024055.

Outay, Fatma, Hanan Abdullah Mengash, and Muhammad Adnan. "Applications of unmanned aerial vehicle (UAV) in road safety, traffic and highway infrastructure management: Recent advances and challenges." *Transportation Research Part A: Policy and Practice* 141 (2020): 116–129.

Zhang, Chunsun. "An UAV-based photogrammetric mapping system for road condition assessment." *The International Archives of the Photogrammetry, Remote Sensing and Spatial Information Sciences* 37 (2008): 627–632.

24 Facial Recognition
Translating Algorithmic Promise into Safer, More Sustainable Indian Cities

Ayushi

ABSTRACT

This chapter explores the transformative role of facial recognition (FR) technology powered by advanced AI and data science in creating safer, more sustainable urban environments across diverse Indian cities. Through detailed case studies of Surat, Delhi, and Chennai, it highlights FR's capacity to address complex urban challenges – ranging from real-time crime detection and missing persons identification to predictive enforcement of repeat offenders. The discussion emphasizes robust system architectures; human-centered design principles that ensure transparency, privacy, and trust; and ethical governance while aligning these technological interventions in FR deployments with the Sustainable Development Goals (SDGs). The chapter offers insights on scalable adoption pathways and measurable impacts, providing a blueprint for policymakers and technologists worldwide to harness FR for inclusive and resilient smart cities.

24.1 INTRODUCTION

Rapid urban growth, dense transport hubs, large informal economies, and floating populations complicate policing and public service delivery. In this crucible, small failures scale quickly: a missed lead in an investigation, a crowded platform without crowd-control measures, or inefficient traffic enforcement multiply into social and economic costs.

India, home to more than 1.4 billion people and a dynamic economy, is one such nation. Each step India takes toward the Sustainable Development Goals (SDGs) ripples outward, shaping the global future. But how does a country as vast and varied as India move toward these shared goals – the world pledged in September 2015 to pursue the United Nations' SDGs – especially in its ever-expanding cities? The answer, increasingly, lies in smart, data-powered solutions. Perhaps nowhere is this more visible – and more misunderstood – than in the city-wide adoption of facial recognition technology (FR).

FR technology has moved beyond a niche policing tool into a mainstream city instrument – not only to deter and solve crime but to improve emergency response, manage crowds, and optimize services. This chapter blends case studies with personal insights, drawing from three Indian cities that tackled radically different urban problems with FR. It demonstrates how powerful data-driven tools can directly support sustainability, inclusion, and safety.

DOI: 10.1201/9781003487531-30

24.2 TECHNOLOGY FOUNDATIONS – HOW DATA SCIENCE POWERS FACIAL RECOGNITION

Imagine walking into a railway station at dusk. The crowd surges, faces blur. Your eyes dart, seeking a familiar face. You recognize your friend not through conscious effort but because your brain acts as a natural pattern-matching engine, trained by thousands of encounters over time. Facial recognition systems mirror this human ingenuity – but with a scientific backbone.

- **Data Gathering:** Surveillance cameras in city streets, transport hubs, and festivals stream vast rivers of video into FR servers. Each frame becomes a mini-puzzle: locating faces amid shifting light, dense crowds, or painted festival attire.
- **Turning Faces into Data:** FR systems translate the human face into a mathematical object – a "faceprint." This process unfolds in stages:
- Detection: Identifying the presence of a face in an image.
- Alignment and Normalization: Because faces rarely appear neatly front-facing, the system virtually rotates and scales each one to a standard orientation.
- Feature Extraction/Embeddings: Converting face geometry into a fixed-length numeric code.
- Matching and Thresholding: Comparing this code against millions of stored templates. Each comparison produces a confidence score, and if it clears a quality threshold, the system flags a match.

Underlying this process are deep neural networks. Much like the human brain learns to recognize a friend's smile after repeated encounters, these models improve with every example they analyze – learning subtle cues as systems retrain, refine, and recalibrate to evolving conditions.

24.3 FACIAL RECOGNITION FOR SOCIAL GOOD – WHICH SDGS ARE IMPACTED AND HOW

Facial recognition systems in cities intersect directly with several SDG targets:

SDG 11 – Sustainable Cities and Communities: FR improves safety in public spaces, enables resilient emergency responses, and helps planners understand mobility patterns for better transport and land-use decisions.

SDG 16 – Peace, Justice, and Strong Institutions: Biometric evidence accelerates investigations, improves case resolution, and strengthens institutional capacity to maintain rule of law.

SDG 13 – Climate Action: FR's efficiency also means fewer wasted resources – linking leaner operations indirectly to climate action goals.

24.4 DIVERSITY OF INDIAN CITIES – TAILORED SOLUTIONS

India's diverse cities teach us that there's no "one-size-fits-all" approach. FR's value emerges when applied to a specific local challenge.

- Surat needed to manage transient populations in its diamond-driven economy.
- Delhi confronted the challenge of reuniting missing persons in a city of vast migratory flows.
- Chennai focused on keeping repeat offenders in check through real-time criminal database matching.

24.5 DIAMONDS AND DATA: SECURING SURAT THROUGH TECHNOLOGY

Context: Surat, India's diamond capital, faced rising safety challenges due to rapid urban growth, a population of over 5 million, and a large influx of transient visitors. High commercial footfall and transient visitors created a policing gap. Traditional policing struggled with thefts, vandalism, and the difficulty of monitoring such a dynamic city in real time.

Solution Journey: The city introduced an integrated surveillance system combining high-resolution cameras, AI video analytics, and facial recognition. Live feeds streamed into a centralized command center, where real-time alerts improved situational awareness. Facial recognition matched individuals against law enforcement databases, aiding both instant response and forensic investigations.

Experience: What struck me most was not just the tech, but its social ripple. Visiting the control room, you can see how technology empowered officers with quick alerts and clear visuals. Citizens and business owners shared feeling safer, proving that with thoughtful design, technology genuinely enhances everyday urban life.

Impact: Within a year of deployment, Surat reported a 27% reduction in crime and over 150 cases resolved efficiently through FR-enabled rapid identification and police response.

24.6 FINDING THE MISSING: FACES IN THE CROWD OF THE CAPITAL CITY DELHI

Context: Delhi, India's capital and one of the world's largest urban agglomerations, faces a unique policing challenge: the high number of missing persons cases, especially women and children. Each year, thousands of families file reports, but traditional methods of manual matching across massive records often left cases unresolved.

Solution Journey: Delhi implemented advanced FR-enabled surveillance across transport hubs and public spaces to identify persons of interest, missing individuals, and monitor crowd behavior during large-scale events. The system was designed to recognize faces even in varied lighting and with partial coverings – as commonly seen during cultural festivals.

Experience: Technology was woven seamlessly into Delhi's fabric. Tasks that once took officers weeks or months could now be completed in minutes, with the system flagging high-probability matches for human verification. This workflow didn't replace human judgment but amplified it, making investigations faster and more precise.

Impact: Within the first few weeks of adoption, the FR system helped authorities identify over 2,900 missing children, with a success rate unmatched by manual processes. Families described the technology as a lifeline that restored hope, turning what once took months into a matter of days or even hours.

24.7 EYES ON THE ELUSIVE: CRIMINAL IDENTIFICATION IN CHENNAI

Context: Chennai, a bustling metropolis on India's southeastern coast, faces the classic challenge of a fast-growing urban hub: keeping track of repeat offenders and ensuring that law enforcement stays a step ahead. Traditional policing relied heavily on manual checks, eyewitness accounts, and reactive measures – often leaving officers one step behind in fast-moving criminal cases.

Solution Journey: Chennai deployed a system that married its extensive live CCTV feeds with facial recognition linked to criminal databases. Instead of relying on after-the-fact reviews, the system scanned faces in real time, instantly cross-referencing them with a centralized watchlist of known offenders. When a suspect entered a monitored area, the system flagged officers instantly, allowing swift interception and deterrence.

Experience: The design prioritized speed and discretion – alerts were routed quietly to officers' handheld devices, minimizing public panic while enabling swift response.

Impact: Officers reported that FR shifted policing from reactive chases to proactive prevention, enhancing their confidence and efficiency on the ground. Chennai police increased timely detection and interdiction of repeat offenders, contributing to measurable reductions in street crime in targeted districts.

24.8 DESIGN THINKING AND HUMAN-CENTERED APPROACH: HOW IT ELEVATES SUSTAINABILITY

FR systems should be designed not as surveillance monoliths but as adaptive, empathetic, and context-sensitive tools that serve people. The goal is not just recognition but resilience – helping cities deliver safety, inclusion, and sustainability. Design thinking brings this to life by centering the end-user, asking: Who benefits? What problem are we solving? How do we ensure equity, trust, and long-term impact? In practice, this means embedding empathy, transparency, and accountability into every stage of deployment.

24.9 PRINCIPLES OF HUMAN-CENTERED FACIAL RECOGNITION

1. **Trust through transparency**: Citizens must understand what FR is being used for (e.g., finding missing children) and what it is *not* being used for. Clear communication reframes it from "watching" to "protecting."
2. **Privacy safeguards**: Aggregated analytics, anonymized storage, and strict access protocols ensure the system protects individuals while still enabling insights.
3. **Human-in-the-loop oversight**: Human oversight ensures technology supports rather than replacing decision-making, maintaining accountability. It delivers relevant, high-confidence notifications to authorities, preventing alert fatigue, and maintains accountability. Technology should augment, not replace, human judgment – final decisions rest with trained officers.
4. **Community engagement**: Awareness campaigns shift the narrative from "watching" to "protecting." Public-facing communication campaigns clarify purposes, such as aiding in finding missing children, shifting perceptions from surveillance to assistance.

24.10 BUILDING BLOCKS OF FACIAL RECOGNITION SYSTEMS

Facial recognition is a carefully layered system, an orchestrated ecosystem of hardware, data, algorithms, and human workflows. For cities to truly benefit from FR, it helps to break down its technical building blocks:

- Cameras as Data Streams: CCTV units at intersections, transit hubs, and public venues act as the "eyes," constantly capturing visual information.
- Processing Pipeline: Edge computing enables near-instant detection where latency is critical, while centralized servers allow large-scale forensic analysis. Cities must calibrate the mix based on urgency, bandwidth, and budget. Choosing the right balance is a core policy decision.
- Databases: Known offenders, missing persons, or persons of interest form the "memory" of the system. FR only works if it has something to match against: criminal records and mugshot databases for law enforcement, missing persons registries for social services and family reunification, and watchlists for event security or counterterrorism. Governance matters here – who maintains these databases, how often they're updated, and how access is logged determine both accuracy and public trust.
- Algorithms: The pattern recognition engine powered by neural networks transforms faces into faceprints and calculates similarity scores. Deep learning models generate face

embeddings and similarity scores, but their value depends on calibration, retraining, and bias mitigation.

- Integration and Action Layer: Detection is only half the story. Command and control centers serve as the "brain," routing feeds to large visualization walls and operator consoles. Dashboards in command centers, discreet mobile alerts for officers, APIs for case-management software help in the integration and subsequent action. Without this translation layer, insights stay trapped in screens instead of guiding action. The quality of this integration determines whether FR insights stay trapped on a screen or translate into faster, safer interventions.
- Deployment Models: Cities choose on-premise, cloud, or hybrid architectures. Each has trade-offs between data control, scalability, and compliance with privacy laws.

Together, these building blocks create an end-to-end ecosystem where data science directly supports the mission of safer, more resilient cities.

24.11 ADOPTION PATHWAYS FOR CITIES

Cities don't adopt FR overnight. Successful pathways often follow three stages:

- Pilots in high-value zones (airports, transit hubs) test feasibility and accuracy.
- Institutional integration connects FR with command centers, existing CCTV, and investigative workflows.
- Scaling with governance introduces independent audits, citizen feedback, and performance benchmarks to preserve trust as the system expands.

These pathways ensure FR grows organically, adapting to each city's rhythm rather than being imposed top-down.

24.12 TANGIBLE IMPACTS ON THE GROUND

The real power of FR lies in what it enables on the ground:

- Force multiplier effect: One officer with real-time FR alerts can do the work of many, shifting policing from reactive to proactive.
- Performance benchmarks: Performance matters as much as deployment. Measurable improvements – faster case resolutions, reduced response times, safer public spaces – turn technology into accountability. Independent bodies like the U.S. National Institute of Standards and Technology (NIST) regularly test accuracy, speed, and resilience under real-world conditions such as poor lighting, occlusions, and aging faces.
- Everyday reassurance: Citizens, commuters, and business owners report feeling safer when they know the city is not just watched but cared for intelligently.

Taken together, these design principles, technical building blocks, adoption pathways, and tangible impacts form a blueprint for sustainable urban safety. While drawn from Indian cities, the lessons resonate globally. Its effectiveness comes from how the building blocks interlock and how well cities design adoption pathways.

24.13 BEYOND RECOGNITION: THE FUTURE OF SAFER CITIES

In the march toward the SDGs, each face in the crowd is more than a data point – it's a story, a possibility, and a promise for a smarter, more sustainable tomorrow. The lessons from India can inform any city embracing FR for SDG-driven improvements. When adopted globally, the principle of human-centered design in FR – anonymized analytics, smart alerting, human judgment, and proactive public dialogue – becomes the cornerstone of ethical and effective urban safety solutions. By integrating these principles, policymakers and technologists worldwide can create urban safety solutions that respect citizens' rights while maximizing operational efficiency. In this model, FR stops being a surveillance story and becomes a civic service story – one where data science delivers measurable public good.

BIBLIOGRAPHY

Apene, Oghenevovwero Zion, Nachamada Vachaku Blamah, and Gilbert Imuetinyan Osaze Aimufua. "Advancements in crime prevention and detection: From traditional approaches to artificial intelligence solutions." *European Journal of Applied Science, Engineering and Technology* 2, no. 2 (2024): 285–297.

Llauradó, José M., Francisco A. Pujol, David Tomás, Anna Visvizi, and Mar Pujol. "Study of image sensors for enhanced face recognition at a distance in the smart city context." *Scientific Reports* 13, no. 1 (2023): 14713.

Musthafa, A. Syed, S. Dinesh, Dinesh Kumar, C. Jeeva, and S. Madesh. "Digital vigilance: AI solutions in the quest for missing persons using face recognition with deep learning algorithms." In *2024 2nd International Conference on Artificial Intelligence and Machine Learning Applications Theme: Healthcare and Internet of Things (AIMLA)*, pp. 1–6. IEEE, 2024.

Ponmalar, A., B. Sandhiya, M. Bhuvaneswari, M. Gayathri, G. K. R. Bhavana, and S. Aarthi. "Finding missing person using artificial intelligence." In *2022 International Conference on Computer, Power and Communications (ICCPC)*, pp. 562–565. IEEE, 2022.

Wang, Xukang, Ying Cheng Wu, Mengjie Zhou, and Hongpeng Fu. "Beyond surveillance: Privacy, ethics, and regulations in face recognition technology." *Frontiers in Big Data* 7 (2024): 1337465.

25 Revamping City Surveillance through Deep Learning and Artificial Intelligence

Juzer Bharmal and Gayatri Srinivasan

ABSTRACT

An intelligent traffic management system (ITMS) was developed and implemented in the city of Ahmedabad to control unruly traffic and minimise traffic violations at major junctions in the city. AI-based cameras were fitted upon over 60 traffic junctions to detect traffic violations. The camera would snap evidence images at high speed, automatically pick up the number plate, access the vehicular database, and identify vehicle owners. The challan issuing process was automated as well.

The ITMS generates over 1 lakh e-challans a day for over 60 junctions, with police dispatching at least 30,000 challans (15 times more challans) every day. This has led to better discipline on the roads of Ahmedabad. Other benefits include tracking lost or stolen vehicular data and tracking vehicles that are involved in criminal activities. In this way, artificial intelligence supports the police in maintaining order in the city.

25.1 INTRODUCTION

Violations of traffic rules are prominent in India, especially across the bigger cities. As road accidents rise in alarming numbers, the primary reason for their occurrence remains traffic violations. These can range from wrong side driving to speeding and running traffic lights.

One of the most frequently observed traffic violations pertains to running traffic lights. This not only increases the probability of accidents but also tends to cause obstructions and chaos at traffic junctions. This violation of traffic rules causes massive loss of life and property. In Ahmedabad, traffic rule violations were rampant because traffic police officers were unable to challan most rule violators and impose fines using the existing information traffic management system (ITMS).

Thus, traffic police were facing huge problems in enforcing traffic rules and instilling discipline on the roads in Ahmedabad effectively and efficiently. How does data science solve this problem?

One of the main reasons that enforcing traffic rules on the roads of Ahmedabad can be very challenging is due to the sheer volume of offenders as compared to the number of traffic police officers that can be deployed in the field to manage traffic.

In the current system, a huge number of challans needed to be issued manually on a daily basis for this intention. Its management necessitates a comparably large task force, which is neither practically feasible nor cost effective to sustain in the long-term.

The practice of manual challan was present in the previous City Surveillance Intelligent Traffic Management System (CSITMS) project across 80 junctions. This system yielded a total of approximately 2000 challans a day, translating to only 25 per junction per day, which is miniscule compared to the real-time traffic offenders crossing each junction per day. The CSITMS was managed by a team of 20 officers working in shifts.

DOI: 10.1201/9781003487531-31

25.2 CURRENT SCENARIO

There were several limitations in the existing system. A significant difference was found between the actual number of traffic violations and the number of challans generated by CSITMS per junction per day. It meant that the challaning system was inefficient since it was being done manually by traffic police officials based on the monitoring of video feeds recorded by the ITMS.

The effort in turn translated into the overall expense involved in issuing a challan, which was much higher. This also rendered the existing process of issuing challans ineffective from a cost perspective. Furthermore, the existing process was also not serving its ultimate purpose of controlling and reducing traffic violations on roads by way of finding violators and issuing challans as a deterrent.

25.3 SOLUTION

Lookman Electroplast Industries Ltd., together with Ahmedabad Municipal Corporation, addressed this issue by deploying a new AI-based ITMS, operating across over 60 junctions. This ITMS requires a much smaller team of officers to monitor it and issues at least 12–15 times more challans in a single day by partially processing high-accuracy automatically generated e-challans. To put this into perspective, the current ITMS issues almost 1 lakh e-challans generated per day compared to 2000 challans issued per day earlier.

Vehicle numbers are caught by cameras deployed by the ITMS which read vehicle numbers of offending vehicles through optical character recognition (OCR) technology. Once the number is captured, it is run through the VAHAN database of the traffic authorities, and vehicle owners are identified. VAHAN is India's centralized, digital national vehicle registry under the Ministry of Road Transport and Highways. Then, challans are automatically prepared and provided to the traffic authorities for necessary action.

This accurate challaning system along with evidence images has greatly increased commuter discipline on the roads of Ahmedabad. This leads to improved traffic control and management at several previously congested junctions in Ahmedabad.

How was this system designed?

At the outset, all violation detection is automated using AI-based cameras. The camera identified real-time traffic offenders. Next, the entire challan generation process is automated, doing away with any manual effort to generate each challan, which was earlier done by an official.

Other aspects like automatic number plate recognition (ANPR), data collection, violation detection with evidence images, accessing the VAHAN database to get driver details based on license plate number detecting, and generating an e-challan were also fully automated to strengthen the process.

Once a challan is generated, an approval life cycle for the challan (i.e., approve, discard, regenerate, or print) is incorporated so that a traffic police official can supervise and verify each challan, make corrections if needed, and print the final challan to then be dispatched to the violators.

Further, an option for manual generation of a challan has been retained in the system to ensure backward compatibility and to store information of challans that are issued by the traffic police official on the spot in person.

25.3.1 Data Used for Solution and Challenges in Data Procurement

The most potent challenge in data procurement was in collecting and automatically analysing video streaming data obtained from CCTV cameras in real time. The system was implemented in 40 different junctions at the time of going live, and currently, it is live across 60 junctions.

The system observes the video feed in real time, implements AI-based automatic number plate recognition for all vehicles passing through all junctions, and detects two different types of violations – red light violations and bus rapid transit system (BRTS) lane violations, which means capturing cars and other vehicles that are wrongly using the lane reserved only for buses.

The new database management system also had to be easily scalable because more junctions were going to be added in the near future. The number of violations detected was also going to increase to include other types of violations such as no helmet, triple riding, and speeding. This required large database storage, often up to 1 Terabyte (TB) per day.

It is important to note here that a separate artificial intelligence model has to be developed for each type of traffic violation type, as the signature of each violation is different. For example, the offending signature of "not wearing helmet" is different from that of "speeding vehicle". The CCTV camera and connected computer cannot automatically understand whether a certain traffic movement is a violation of the traffic law. Hence, the AI algorithm was specifically trained to recognise each traffic violation. The question is: How is this done? Using the previous example, for the violation of "not wearing helmet", as described, the AI algorithm was fed several images where the person on the bike was wearing a helmet and then a large number of images where the person was not wearing a helmet (traffic violation case). Based on these images, the AI algorithm develops a mathematical model at the backend which can differentiate between a violation scenario and a legitimate scenario. Further, when the system is deployed, this trained mathematical model is used to identify traffic violations in real time and generate traffic challans.

This is not a small feat. Compare the scenario of a major traffic junction that witnesses lakhs of vehicles every single day. No matter how many traffic policemen are deployed to handle such a junction, they will not be able to catch violating vehicles, impose penalties, and therefore discipline the traffic the way the ITMS has been able to.

Gradually, the system started being used to track more offenders. For example, a stolen/wanted vehicle alert was implemented, which meant capturing data of over tens of thousands of vehicles passing through all the junctions per second. The system would automatically detect number plates, then access the central database server with about 1 lakh stolen/wanted vehicles stored in it, cross-check if a blacklisted vehicle was matched with the real-time traffic data, and then sound an instantaneous pop-up alert in the dashboard if such a vehicle was found. All these transactions are completed within the span of just 1 second.

The second challenge was equipping the system so that it could search through the database based on various filter criteria such as date range, license plate number, junction name, and pole name. These details are provided along with search results processed and need to be displayed within a matter of seconds.

In addition, daily Management Information System (MIS) reports were also necessary for the police to analyse junctions witnessing maximum violations, the number of top violators, and preparing challan summaries.

To address these challenges, we chose a big data solution which was implemented using MongoDB, an open source database management system (DBMS). This uses a document-oriented database model which supports various forms of data. It gives users the flexibility to create any number of fields in a document, which was a primary need for this solution.

Further, the data in a MongoDB collection can be distributed across multiple systems for horizontal scalability to keep up with data volume and increasing throughput requirements. Also, the solution lets users connect to cloud services and set real-time triggers in databases.

25.3.2 Challenges in Development

Usually, good-quality images are required for the computer algorithm to detect traffic violations. But since the vehicles are moving, often images taken by the CCTV camera are not clear, and thus post-processing has to be carried out for the computer to identify the vehicle and detect any violation. In some scenarios, only a partial image is captured, and in such scenarios violation detection cannot be done by the computer algorithm due to insufficient information. A lot of times, video readings are not perfect. A lot of outside factors can give rise to partial number plates. Therefore, the system has to undertake a partial string search. This means the system searches based on a partially recorded license plate number and matches it with other important factors such as make, manufacturer, brand, and colour of the car to determine the partial number with the exact vehicle.

Another challenge was that of data maintenance. The system maintains crores of data points recorded at a very fast speed in real time. Therefore, server processing capacity was also an additional challenge faced by the development team.

The system is also built to analyse patterns in the data being recorded, such as patterns of violations across junctions. These are then used to create statistical MIS reports, which were also challenging. However, the system needed higher accuracy for classification.

From an analytics perspective, in traditional systems, only 60% accuracy is achievable for license plate recognition, specifically factoring in Indian roads' conditions. This is because of tailgating vehicles that provide too few frames of the vehicle number plate. Tailgating vehicles also challenge the OCR system owing to the different fonts and formats used in Indian number plates, non-standard number plates, or damaged number plates affixed upon vehicles operating on roads.

The accuracy of such systems improves over time once better training data is available. Manual inspection and continuous quality checks have to be carried out to ensure the computer algorithm is working properly and to debug any issues or glitches.

In this scenario, the team used AI-based deep learning technology to improve the accuracy of the system. This technology has improved progressively with training, which has resulted in over 90% accuracy in license plate recognition. The analytics software has also demonstrated a cognitive ability to recognise letters, which is difficult even for humans in actual conditions.

25.4 IMPACT

The AI-based ITMS has been a huge success in traffic management at Ahmedabad. First, it has dramatically reduced the time taken to generate a challan. From only 2000 challans generated daily for 80 junctions, the system now automatically generates over 1 lakh e-challans per day for over 60

FIGURE 25.1 A typical junction in Ahmedabad prior to implementation of the red-light violation system.

FIGURE 25.2 The same junction post-implementation of red-light violation system in the year 2018.

junctions. The police now dispatch at least 30,000 challans every day with the help of the AI-based ITMS. This figure is over 15 times the number of challans issued earlier. Therefore, the biggest benefit of the system has been its sheer output using less human labour and human hours.

The cascading effect, which was the core agenda behind developing and bringing the system in place, is also materialising. There is a visible increase in discipline on the road and sharp reduction in stop line violation.

By deploying the new AI-based ITMS, currently running across over 60 junctions, a much smaller team of officers easily issues at least 12–15 times more challans in a single day after partially processing high-accuracy automatically generated e-challans.

The accurate challaning system, along with evidence images, has increased commuter discipline on the roads of Ahmedabad, leading to better traffic control and contributing to overall road safety.

BIBLIOGRAPHY

Anand, S. Raj, Naveen Kilari, D. Udaya Suriya Raj Kumar, and Sai Nitisha Tadiboina. “Traffic signal violation detection using artificial intelligence and deep learning.” *International Journal of Advanced Research in Engineering and Technology (IJARET)* 12, no. 2 (2021): 207–217.

Charran, R. Shree, and Rahul Kumar Dubey. “Two-wheeler vehicle traffic violations detection and automated ticketing for Indian road scenario.” *IEEE Transactions on Intelligent Transportation Systems* 23, no. 11 (2022): 22002–22007.

Franklin, Ruben J. “Traffic signal violation detection using artificial intelligence and deep learning.” In *2020 5th International Conference on Communication and Electronics Systems (ICCES)*, pp. 839–844. IEEE, 2020.

Said, Yahia, Yahya Alassaf, Refka Ghodhbani, Yazan Ahmad Alsariera, Taoufik Saidani, Olfa Ben Rhaiem, Mohamad Khaled Makhdoum, and Manel Hleili. “AI-based helmet violation detection for traffic management system.” *Computer Modeling in Engineering & Sciences (CMES)* 141, no. 1 (2024).

26 Location Casting
Identifying Demand Clusters via Big Data

Akhilesh Tilotia[1]

ABSTRACT

When the Ministry of Civil Aviation (MoCA), Government of India, was launching its *Ude Desh ka Aam Naagrik* (UDAN) scheme, it wanted to know where new airports should be built. The UDAN scheme required airlines to bid for routes connecting airports that were unserved or underserved. While the industry did its own demand estimation, MoCA wanted to understand where it could expect demand. MoCA took a big data approach, working closely with a team at Google.

The project identified clusters where there was a willingness to travel coupled with the ability to pay. Google dove deep into its databases of searches and locations, meshing them with machine learning algorithms and artificial intelligence (yes, in 2017) to define cities/clusters where the data suggested there could be viable demand for an airport. However, an airport needs to be seen in context of the entire aviation network, requiring further parsing of data to balance the conflicting objectives of profitability of airports and accessibility for citizens concurrently.

How well did this work out? Of the ten cities identified (which were not shared with the industry), seven were picked by the industry in the first round of UDAN bidding as places from which they wanted to start flights. In the next round of UDAN bidding, 20 of the 30 cities identified were picked by the industry.

India's vastness, massive population, and rising wealth have long been at odds with the country's underdeveloped aviation infrastructure. As of 2016, just 75 airports were operational across the country, a figure that was woefully insufficient given the size and scale of India's domestic travel needs. The launch of the Ude Desh ka Aam Naagrik (UDAN) scheme, designed to boost regional connectivity, marked a turning point. Before this initiative, the task of traveling from one end of India to the other could take up to 48 hours by train, a journey that could be completed in a mere three hours by air. With estimates suggesting that between 10 and 30 million unique individuals had flown in India, the opportunity to connect the remaining 97% of the population – who had yet to experience air travel – was apparent.

The growing middle class, combined with a steady decline in ticket prices, further fueled demand. Air travel had become more accessible, with the average cost to passengers standing at just Rs 4 per kilometer. At roughly six cents per kilometer, it was about half the global average. More importantly, this was broadly similar to the cost of travelling by an air-conditioned coach in Indian Railways. By FY2018, hence, domestic air ticket sales were almost equal to the total number of air-conditioned train tickets sold by Indian Railways, with around 120 million domestic flight passengers compared to 130 million air-conditioned rail tickets. This shift towards air travel highlights not just a rise in affluence but a profound change in how India moves.

DOI: 10.1201/9781003487531-32

26.1 THE TRADITIONAL APPROACH: A DATA-STARVED ESTIMATE OF DEMAND

Historically, determining the location of new economic ventures – whether airports, retail outlets, or entertainment venues – relied on a mix of ground surveys and top-down assessments of regional affluence. On-site surveys would gauge demand proxies, querying a sample of potential customers on their willingness to spend. However, the accuracy of such forecasts hinged not just on willingness but also on the ability to pay. This complex relationship often led businesses to consult economists and statisticians to parse through proxies indicating purchasing power. In India, the available data typically provided district-wise per capita GDP figures, but such broad strokes were hardly granular enough to offer meaningful insights at the city level. In practice, this process often involved numerous assumptions, approximations, and outdated analysis, leaving much to be desired in terms of precision.

26.2 THE BIG DATA SHIFT: A MORE PRECISE APPROACH TO FORECASTING DEMAND

In the age of big data, India's Ministry of Civil Aviation (MoCA) wondered whether it was possible to harness the massive data output generated by the country's digital activity to more accurately assess demand for air travel. India, now one of the world's largest consumers of data, were consuming over 5.5 billion GB of data by the fourth quarter of 2017 alone, and it ranked second globally for mobile search queries. Could this vast digital footprint hold the key to identifying patterns of travel demand and, more importantly, the ability to pay?

MoCA partnered with Google to pose an intriguing question: Could data-driven insights be used to pinpoint potential locations for new airports? With hundreds of millions of search queries generated daily, many related to travel, there was ample information to work with. Indians regularly searched for routes, comparing travel times, costs, and modes of transport, as well as seeking details about safety and comfort. The vast scope of these queries offered an unprecedented opportunity to move beyond assumptions and approximate metrics and instead mine actionable data that could highlight where demand for air travel – coupled with the ability to afford it – was most concentrated.

This new approach coincided with the government's ambitious launch of the Ude Desh ka Aam Naagrik regional connectivity scheme, aimed at making air travel more accessible across India. Leveraging industry support (which included waivers for landing, parking, and navigation charges), local government backing (including tax incentives and significant local support), and market-friendly regulations (such as three-year exclusivity on routes and viability gap funding), the UDAN initiative has proved a resounding success. From just 75 operational airports at the start of 2016, India had close to 140 operational airports by the end of 2018. Figure 26.1 shows the initial list of airports that were operationalized under the UDAN connectivity scheme. A complete list of all the airports, along with the date of starting operations, can be found on the Ministry of Civil Aviation portal.[2]

26.3 THREE ASPECTS TO CONSIDER

Three things were important for the project: demand, network optimization, and ability to refresh.

The basic definition of demand, as different from a desire, is that it adds a layer of ability to pay on the willingness to consume. The project needed a way to identify not just where the desire to travel existed but where there was an ability to pay – this required creating two layers and then identifying the common points of interest.

After the promising locations for new airports were identified, it was important to consider not just the location of the individual airport but the overall network. Largely aviation works on a "hub-and-spoke" model, and customers are willing to travel a certain distance to access the aviation infrastructure. The outcomes needed to trade off between creating large infrastructure assets and their utilization.

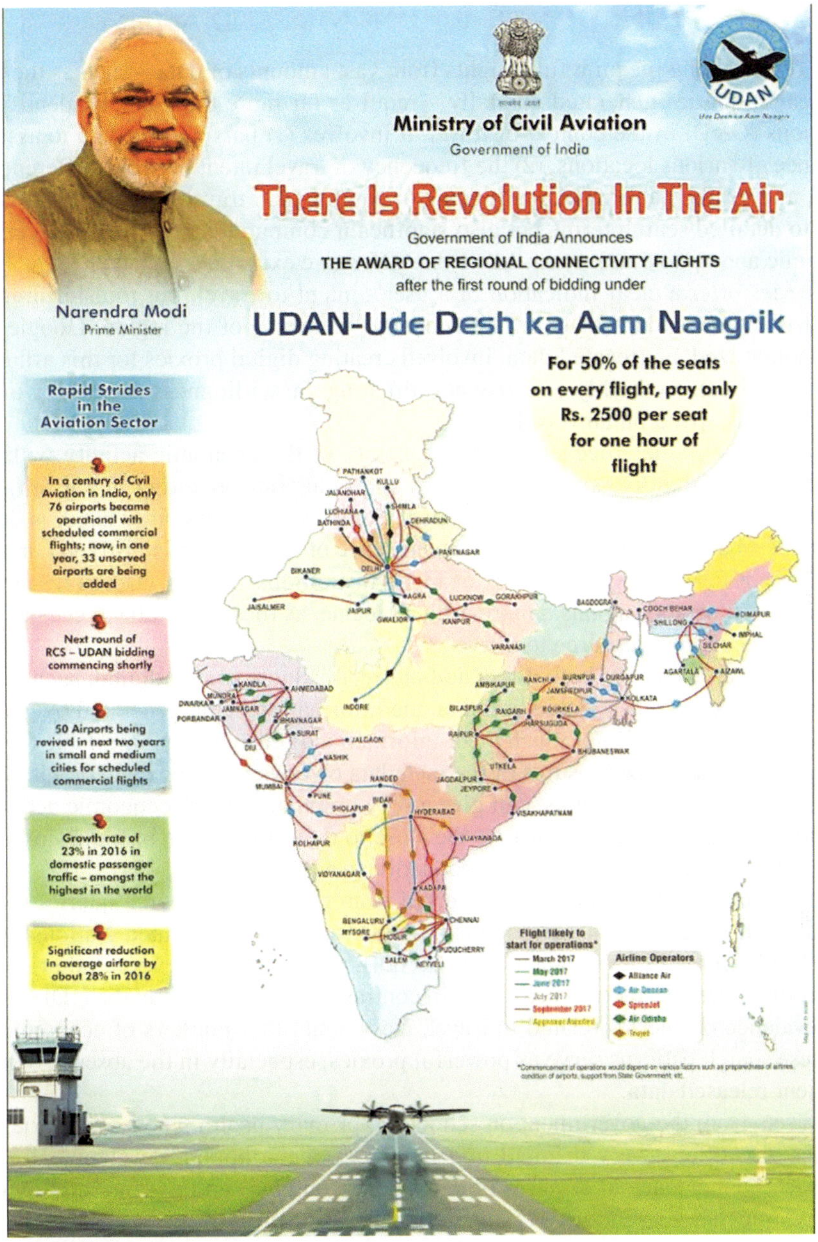

FIGURE 26.1 Initial list of airports operationalized under the UDAN scheme.

These two points are in a sense *de rigueur* for any analysis, even in the world of static, one-time analysis before an airport is commissioned. One key innovation in this method was that the underlying information powering the model would continue to evolve as the customer trends, willingness, and ability to pay evolved. India is a fast-growing economy – a 6–7% real growth per annum means that the size of the real economy doubles in around a decade. Aviation is a "tipping point" industry: once a location reaches a reasonable level of affluence, consistent demand can suddenly surface; alternatively, if a large industry (manufacturing or services) hubs around a particular location, there can be a spike in demand for travel to and from there.

26.4 DEMAND: A FUNCTION OF WILLINGNESS AND ABILITY

The process of extracting meaningful insights from vast amounts of data – such as the hundreds of billions of search queries generated annually – requires far more than simply identifying origins and destinations (O&D) in the context of travel. It involves (1) parsing this data to assess the relative importance of various locations, (2) the frequency of travel intent between different origins and destinations, and (3) the patterns of willingness to pay for such travel. Such analysis demands not only access to detailed search terms but also significant computational power, especially if it is to remain dynamic and up to date, rather than a one-off, static exercise.

Search queries offer a clear indication of a user's intent to travel, but transforming that intent into actual demand depends heavily on the economic affluence of the region. Google's approach, leveraging anonymized, aggregated data, involved creating digital proxies for this affluence. These proxies, while indirect, are highly effective at estimating the willingness and ability of consumers in various regions to spend on air travel.

One key proxy is the heterogeneity and complexity of the economic activity within a region. Heterogeneity reflects the diversity of businesses and industries operating in a given area – more variety generally correlates with higher affluence. Complexity, on the other hand, is driven by the interdependencies between businesses or by the presence of specialized industries. For instance, the presence of freight-forwarding companies or niche institutions such as fashion design schools or specialized research centers signals a more affluent locale, as these specialized ecosystems require a certain level of economic resource to thrive.

In India now where taxation, both direct and indirect, is increasingly digital, such heterogeneity and complexity can also be calculated based on the information being generated by goods and services tax data or income tax filings. Such meta-information is currently not available for the analytical community to work on. Disclosures of the metadata on e-way bills or input credit, coupled with the cluster of location of transactions, can give a very granular view of economic activity in India. The data from FasTag toll collection across the close to thousand plazas in India similarly offers a good view of the texture and granularity of economic growth.

Moreover, human development can be inferred from the types of search queries prevalent in a region. Areas with lower human development, as measured by the United Nations Development Program (UNDP) Human Development Index (HDI), tend to generate more adult-related search queries, reflecting different societal priorities. In contrast, regions higher on the HDI scale often see a greater prevalence of queries related to travel, news, and other markers of economic and social mobility. These search patterns serve as powerful proxies, especially in the absence of more granular, government-released data.

Large datasets from the government on schooling, primary health, and beneficiaries of government schemes also offer a very detailed understanding of the human development in a locality. Areas with specific incidence of diseases or of academic outputs can be identified as locations ripe for either development intervention or commercial viability. Similarly, locations which increasingly get less dependent on government grants can become important consumption centers, especially for discretionary items. Such datasets, while available in some cases, require deeper and more consistent releases to build robust trendlines.

By combining these proxies – economic diversity, business complexity, and HDI-driven search patterns – Google was able to generate reasonably accurate estimates of air travel demand for cities not yet served by the airport network. This analysis identified at least eight cities that, despite lacking direct air connections, showed independent potential for attracting 500,000 or more passengers annually, confirming the latent demand in these emerging regions.

26.5 OPTIMIZED AIRPORT NETWORK: THE SECOND PIECE OF THE PUZZLE

Understanding demand dynamics is only part of the challenge when it comes to expanding air travel infrastructure. There is a social imperative to providing access to an aviation network in deeper regional and remote areas of the country. It was calculated that the USA, with around 10 million square kilometers of land area, had around 650 airports. India, with one-third the land area had only around 70 active airports then. Using this metric suggested more than tripling the number of airports in India.

India, however, is more densely populated than the USA. Hence, every new airport influences not just its own potential demand but also the broader network's dynamics. Take, for example, Jalandhar and Ludhiana – two cities in Punjab that, from a demand perspective, appear ideal candidates for air connectivity. However, with only 60 kilometers separating them, the presence of two airports so close together could result in demand cannibalization, undermining the profitability of both. Thus, finding the optimal airport network requires simulating how each potential addition would affect overall demand, taking into account the complex interactions within the network.

The real challenge lies in defining what constitutes an "optimal" solution. An airport network must strike a delicate balance between profitability and accessibility – goals that are often in tension. While profitability is straightforward to measure, accessibility is more elusive. To address this, the team used a novel approach to measure "travel overhead" – the extra distance passengers must travel to reach their nearest airport. In India, the average travel overhead per itinerary was 122 kilometers, which underscores the pressing need for more airports in many regions. The project estimated that if 30 new airports were added as identified, then then travel overhead could come down to 77 kilometers, a material reduction that makes access more viable for customers.

These large numbers for travel overhead are significantly influenced by the regional and remote airport connections. The idea behind the UDAN scheme was to reduce this access challenge. For metro city airports, the travel overheads were noted to be much smaller: this ranged between 25 and 35 kilometers for 95% of the passengers at Bengaluru, Delhi, Hyderabad, and Mumbai. However, given traffic congestion in these cities, it was noted that the travel overhead could be measured in terms of number of minutes of commute to the airport and not just the distance.

Quantifying this overhead allows policymakers and planners to make data-driven decisions, optimizing the airport network to not only meet demand but also ensure that air travel remains accessible and efficient for all regions.

26.6 A DYNAMIC NEW WAY OF DECISION MAKING

The integration of willingness and ability to pay for air travel results in a dynamic map of demand opportunities, highlighting where latent demand is waiting to be tapped. This method shifts from traditional, static analysis to a more fluid, data-driven approach where probabilities evolve as new information continually filters in. Unlike the traditional top-down approach, which often relies on outdated, incomplete data without a clear understanding of the likelihood of success, this dynamic model provides a more accurate and real-time picture of market potential.

Initially, MoCA sought to determine the optimal locations for ten new airports. The data-driven recommendations suggested cities that would reduce travel overhead by 20%. Remarkably, seven of the ten cities identified were selected by the industry during the first round of UDAN bidding as prime locations to launch new flights. In the subsequent round of UDAN bidding, 20 out of 30 cities identified through the same data-driven process were picked by the industry for flight operations.

This alignment between the model's recommendations and the industry's choices offers confidence in the efficacy of the data-driven approach. As the model matures and input quality improves, or as a deeper understanding of the relationships between variables is achieved – whether through human insight or machine learning – the quality and accuracy of these forecasts are expected to only improve, further enhancing the strategic decision-making process for India's airport expansion.

One important insight that emerged from the project is the important trade-off between an airport's economic viability (largely based on the number of passengers who access the aviation network using the airport) and the service distance that customers are required to commute. The example of Jalandhar and Ludhiana, when scaled to the level of the country, showcased (then) that the marginal utility of adding airports after 100 operational airports in India may be low. This could change with time as prosperity percolates.

26.7 BROADENING THE SCOPE OF "LOCATION-CASTING": FROM AIRPORTS TO GROCERY STORES

What began as a targeted effort to determine the optimal locations for new airports has evolved into a flexible, data-driven framework with far broader applications. This methodology can now be applied to a wide range of public- and private-sector establishments, from retail outlets to healthcare facilities. One prominent example of this generalized approach in action was its adoption by a leading retailer, which is leveraging the same data-driven principles for market expansion.

The key advantage of this dynamic, continuously evolving approach is that it allows recommendations to be monitored and adjusted in real time as new data and trends emerge. While such agility is invaluable for businesses with shorter timelines and more flexible infrastructure needs, it might not be as practical for long-gestation projects like airports. Nonetheless, for many industries, the ability to make decisions based on "now-casting" – real-time predictive analytics – provides a substantial competitive edge.

For airlines, this framework could be used to identify emerging routes based on shifting demand patterns. Retailers, similarly, can apply it to optimize store placements or refine their existing coverage. In the world of on-demand services, decisions on where to stock inventory at warehouses could be driven by hyperlocal demand signals, ensuring products are available where and when they are most needed. Beyond the private sector, governments can also harness this approach to plan the locations of essential public services like healthcare facilities, schools, and primary care centers, ensuring that resources are optimally distributed to meet community needs.

Despite the power of big data, human judgment will remain essential in interpreting these insights. The real value lies not in replacing decision-making but in empowering leaders with a richer, more informed foundation to guide their choices. By combining the wisdom of experience with the precision of data, businesses and governments alike can make decisions that are both strategic and responsive to the evolving landscape.

NOTES

1 The author served as an officer on special duty with the minister of state for civil aviation (2016–19). It was in this period that the project was conceptualized and executed. This write-up is based on a document that was released as a "Think with Google" paper and the author's own understanding, experience, and recollections. The author is now a co-founder of Thurro, an alternative data platform and analytics company. Views are the author's own.

2 https://www.aai.aero/sites/default/files/rcs_news_notifications/609%20RCS%20Routes%20Operationalised_06.11.2024.pdf

Index

D

E

F

G

H

I

J

K

L

M

9781032781310